Renewable Energy

BRAD LINSCOTT

Renewable Energy

A Common Sense Energy Plan

Tate Publishing & Enterprises

Published by Tate Publishing & Enterprises, LLC
127 E. Trade Center Terrace | Mustang, Oklahoma 73064 USA
1.888.361.9473 | www.tatepublishing.com

Tate Publishing is committed to excellence in the publishing industry. The company reflects the philosophy established by the founders, based on Psalm 68:11,
"The Lord gave the word and great was the company of those who published it."

Published in the United States of America

ISBN: 978-1-61777-608-3
1. Technology & Engineering / Power Resources / General
2. Technology & Engineering / Power Resources / Alternative & Renewable
11.05.19

Dedication

To my children and grandchildren with the hope that they and others will enjoy an improved way of life in our country when we all can say, "We are truly an energy-independent nation!"

Acknowledgments

I am pleased that Tate Publishing selected my manuscript for publication.

I must thank my wife for her patience and understanding during the process of writing this book. I want to convey my appreciation to many friends and former coworkers at the National Aeronautics and Space Administration Research Center in Cleveland, Ohio, for their encouragement and support.

Table of Contents

Foreword

In 2009, I wrote a preliminary energy plan with recommendations for modifying the Department of Energy's renewable energy plan for FY2010. I submitted the plan to my congressional representative and my two United States senators from Ohio. Of the three Ohio representatives, only one—Ohio's United States senator, George Voinovich—responded to my preliminary energy plan.

Senator Voinovich stated in his letter, "It is critical that we grow more energy independent to increase our competitiveness in the global marketplace and improve our national security. As less of our energy needs are met with our resources, our nation is placed at the mercy of the oil-exporting OPEC nations and vulnerable to geopolitical instability and oil market volatility."[1]

My Ohio district United States representative, Betty Sutton, and Ohio's United States senator, Sherrod Brown, did not respond to my inquires, either by letter or e-mail, in 2009. I can only conclude that these two representatives are responding not so much to individuals but probably to special interest groups and lobbyists to maintain political party allegiance and to some political party contributors with deep pockets.

It was this experience that motivated me to write a book on the subject. My hope is that a published book will reach a larger audience and provide sufficient information to convince readers that a change in our energy policy is sorely needed. My desire is to inform our citizens on the broken political promises made by both the Republican and Democratic parties with regard to our energy policies. A revised energy policy must maintain the goals of clean economical energy and energy independence. Perhaps a large group of people with a better understanding of our energy needs can help to sway our country's legislators away from the energy path we are on now.

Introduction

We are told that renewable energy is necessary for "energy independence." We have waited over thirty-five years and paid over $172 billion (the dollar amount is adjusted for inflation in 2005 dollars), and we have not reached the goal set in 1975 of "energy independence." After thirty-five years of spending, the result is that renewable energy provided only 2.5 percent of our energy needs in 2007. This is not an admirable achievement by either our Democratic or Republican administrations over these many years.

During the last few years, energy is back in the political spotlight because gasoline prices have escalated and environmental studies predict the potential for manmade global warming. Gaseous emissions, a by-product of fossil fuel combustion, have the potential to adversely affect world climates. Automotive gasoline engines and fossil-fuel-fired electric power plants emit sufficient quantities of pollution to adversely affect our climate. To reduce the potential of global warming, environmental scientists suggest we must substantially reduce the amount of emissions that result from the combustion of fossil fuels.

How can renewable energy help us to reach this long-sought-after goal of clean energy and energy independence? During the preparation of this book, I have drawn on my thirty years of experience to research, develop, and write technical reports on terrestrial and space energy systems. Many energy related questions have crossed my mind over the last few years. For example, do the objectives of our energy program make common sense? How long must we wait, and how much of our tax money will be spent, before we achieve our goal of energy independence?

Can we depend on renewable energy resources to eliminate our need for imported oil from foreign countries? Is our energy program leading us in the wrong direction? Are our tax dollars for energy being spent wisely? I have found answers to these questions and many others that have transpired during the preparation of this book. It is my desire to share the answers with you. I hope that by your reading this book, you and others, one person at a time, will be better informed about the objectives of our nation's energy program and the associated expenditures needed to support the government's energy program.

The words "one person at a time" remind me of a parable about a large number of starfish that were washed from a storm-driven ocean onto a sandy beach. The day after the storm, the sun was shining on the beach. The starfish were beginning to wither up and die because they were unable to move themselves back to their ocean home. Early in the morning, a young man was walking along the beach near the water. Whenever he saw a starfish, he picked it up and gently tossed the starfish back into the ocean. As the young man continued to toss the starfish, one at a time, back into the water, he met an old man walking toward him. The old man said to him, "Son, there are hundreds of starfish strewn along this beach. What difference can you make?" As the young man picked up another starfish and prepared to toss it into the ocean, he replied to the old man, "It makes a difference to this one."

I was a member of the Wind Energy Project Office at the National Aeronautics and Space Administration's (NASA) Center in Cleveland, Ohio. Direction and funding for wind energy projects was provided to NASA by the Department of Energy. The Wind Energy Project Office managed the design, fabrication, installation of, and research tests on large wind turbines. A primary objective was to demonstrate electrical compatibility of large wind turbines with large and small utility networks in the United States. In addition to wind turbines, I gained valuable experience during the design, fabrication, and test of solar thermal space power systems and during the preliminary design of photovoltaic power systems for earth-orbiting satellites. My thesis, published for the Master of Science degree from Case Institute of Technology in Cleveland, Ohio, analyzed the surface distortions of parabolic solar concentrators subjected to surface temperature variations. Surface distortions adversely affect the reflective efficiency of solar concentrators. For a short period of time, I was a member of the engineering team at the NASA Nuclear Reactor Test Facility located in Sandusky, Ohio.

Part One

Renewable Energy

Our Renewable Energy Resources

What are these so-called renewable energy sources? Five of our renewable energy resources are: wind energy; solar energy; biofuels; hydropower, including ocean tidal and wave action energy; and the fifth, geothermal energy. Wind energy, solar energy, and biofuels are being promoted more than geothermal, hydropower, and ocean tidal/wave energy sources.

Wind Energy

Wind is called a renewable resource because, fortunately for us, at some particular location on the surface of the earth, the wind is blowing. If we could measure wind speeds at every location around the world, over a one-year period we would find that somewhere there is wind blowing every minute of every day of the year. For some reason the word *wind* has been separated from the word *solar*. The word *solar* refers to the utilization of energy derived from the sun. The fact is that wind is a solar phenomenon. Wind is air movement, especially a natural and perceptible movement of air parallel to or along the ground. But why do we have winds? Because during the day, the sun warms the surface of the earth, which in turn warms the surrounding air. During the night, in the absence of the sun, the surface of the earth cools, causing the surrounding air to cool. Variations of air temperatures occur during the daytime and night-time hours. Air temperatures also vary with altitude above ground level. As air temperatures change, the density of air changes, which in turn causes the air pressure to change. The resulting variations in air pressure, caused by solar heating and cooling of the earth's land and oceans, cause the wind to blow. So if we had no sun, we would have no wind.

Wind energy is classified as "green" energy, and it is believed that with enough wind turbines running, they would help to reduce our need for imported oil. But we are not told how many of these machines are needed to gain our "independence" from foreign oil imports.

You have probably seen wind power plants popping up all across our country. The wind turbine manufacturers in Europe and Asia are selling wind turbines for installation around the world. Many wind turbines manufactured in Europe are being purchased by private investors and installed here. Some of our electric utility companies have purchased wind turbines from Europe and from companies in our country for generating electric energy.

Our government is giving our tax money to utility companies and providing tax incentives to private investors for new wind turbines. We are, in effect, paying higher electric rates each time a new wind turbine is installed. Our tax money is also being used to pay workers in European and Asian countries that manufacture and sell wind turbines to our country. Electricity produced by wind turbines costs more than our current cost for electricity. With all of the new wind turbines in service, why haven't our electric bills increased? The reason is because there aren't enough wind turbines operating on utility grids to appreciably cause our electric bills to increase.

Solar Energy

Most of us have heard about the great expectations for "green energy" from newly installed solar-electric plants. The advertising tells us that solar energy is the key to extracting our country from the economic grip of foreign oil imports. The fact is that solar energy remains in an infant stage of development and will not provide a viable economic source of energy during the next ten years. And like wind energy, solar electric power will not provide any significant impact on the amount of oil that we import from foreign countries. Yet our government is spending our tax money to support research and development for solar energy. The total amount of government expenditures for solar energy is difficult to determine because tax payers are subsidizing homeowners and utility companies for the installation and operation of new solar electric systems.

Biofuels

Biofuels are produced from *biomass*, which means recently living organisms, such as various types of plant life. Corn and sugarcane, for example, are commonly used to produce biofuels. These fuels offer the potential to diminish the generation of carbon dioxide that enters our atmosphere when burned in automotive internal combustion engines. Biofuels fuels emit similar amounts of carbon dioxide as gasoline when burned. However, the plant life used to manufacture biofuels have absorbed carbon dioxide during their growing cycle. On the other hand, to grow plants, the soil has to be prepared for planting, seeds have to planted, cultivation and fertilization is often needed, and the plants have to be harvested. The harvested biomass has to be converted to a biofuel, and the biofuel has to be delivered to a distributor. All of these activities require energy from something other than biofuels. The energy sources used to manufacture biofuels are primarily electricity produced by fossil-fuel-fired electric plants and gasoline needed for farming and harvesting plant life used to manufacture biofuels

Is there an actual net reduction of carbon dioxide emissions generated from biofuel combustion? It depends on three factors:

- The amount of carbon dioxide absorbed during the plants' life.

- The amount of carbon dioxide emitted from an energy source needed to grow and harvest the plants and to produce and distribute the biofuel.

- The amount of carbon dioxide produced by the biofuel as a result of combustion.

Our tax money is being used to subsidize farmers to grow corn for the production of a biofuel called "ethanol" under the Energy Independence and Security Act of 2007. The title of the Act is intended to reinforce the perception that biofuels will reduce our dependence on imported oil and provide clean energy sometime in the future.

Hydroelectricity, Ocean Wave, and Tidal Energy

The fourth renewable energy source includes hydroelectric plants, energy derived from ocean wave action, and tidal currents. Hydroelectricity is obtained from water driven turbine-electric generators. The turbines extract energy from the water flowing through them. The water turbines drive electric generators. Hydroelectricity is currently a larger contributor of renewable energy than all of the other reusable energy sources combined. However, over the last few years, the energy provided by hydropower has diminished. A lack of water, due to long-term drought conditions in many of the localities where the hydroelectric plants are located, has caused a reduction in the energy produced by our hydro-plants. The largest sources for hydroelectric power in our country have been developed and are operating. As a result, there are no new locations available for the construction of a large hydroelectric plant. It is anticipated that in the future the percent of renewable hydroelectric energy will diminish compared to the growing need for electricity in our country.

When harnessed, the energy from our oceans can be used to convert wave and tidal energy into electricity. Two primary methods of energy conversion are devices that harness ocean tides and devices that extract energy from ocean waves. There is potentially a tremendous amount of energy available from wave and tidal action. Our ability to extract energy from ocean tides and wave action is at an early stage of researching and developing the technology. Applied research, supported with our tax money, on small-scale devices is being conducted. This effort is intended to help determine feasibility for long-life operation, to determine environmental impact on ocean life, and to estimate the future cost of this type of renewable energy. Like solar energy, ocean wave and tidal action sources for energy production are in the infant stage of development

Geothermal Energy

The fifth type of renewable energy is called geothermal energy. Certain locations in the United States have underground reservoirs of hot water or underground layers of rocks that contain heat energy. These reservoirs can be tapped by drilling through the surface of the earth into hot water reservoirs or hot rock formations. The energy contained in the hot water and hot rock formations can be converted into electricity to supply existing electric utility networks.

Renewable energy sources during 2007 (excluding hydropower) accounted for only 2.5 percent of the total net generation.[2] Of the 2.5 percent contributed by renewable energy sources, wind energy contributed 0.8 percent. Geothermal and biomass energy together contributed another 0.8 percent. The remainder of the 2.5 percent was contributed by wood and its fuel derivatives. The cumulative federal energy expenditures for research and development attributed to just renewable energy and energy efficiency was $26 billion from 1961 to 2008.[3]

A Little Bit of History

Are you tired of hearing, "We must reduce our dependence on foreign oil"? Maybe it's because many of us—especially if you've been referred to as an "old timer"—have heard this declaration since 1974. Many remember the conservation measures imposed on us by our legislators during the 1970s. Interstate speed limits were reduced to fifty-five miles per hour across our nation. Business and government office temperatures were set not to exceed sixty-eight degrees Fahrenheit during the winter months. Homeowners were asked to use fewer, or eliminate the use of, Christmas lights during the holiday season to conserve energy. Homeowners with gas lights used to illuminate their front yards and sidewalks were requested to turn off the gas. If you allowed your car engine to run at idle speed for over fifteen seconds, you were wasting gasoline. The congress authorized tax incentives for homeowners if they would purchase and install insulation in their homes.

Since 2007, when gasoline prices began to increase, we hear again the need to conserve energy. The congress is now providing tax incentives for homeowners to have insulated windows installed in their homes. Our newly appointed director for energy has recently suggested that if we paint all of the roofs in the world with white paint, it will greatly improve our ability to reduce global warming.

Our energy policy changed in 1981, during President Reagan's administration. The Reagan energy policy supported a free energy market that relied on the use of fossil fuels, mainly comprised of gasoline, coal, and natural gas. The policy advocated that our government should not interfere with free market decisions with regard to energy. The contention was that our country could rely on inexpensive, unlimited energy supplies in the future. A brief history of our energy policies is presented, as advocated by each of the Washington administrations, for the last forty years. Since 1981 our energy policies have essentially remained unchanged. As a result, we have made little or no progress toward diminishing our reliance on foreign energy imports.

Our Federal Investment in Research and Development (R&D)

Between 1961 and 2008, our government has cumulatively invested nearly $4 trillion (adjusted to 2005 dollars) in research and development (R&D). For all but one year—1979—defense R&D has accounted for at least half of all federal R&D investments. The space program during the 1960s accounted for as much as one-third of all federal R&D. Beginning in the mid 1990s, health R&D started to account for anywhere between one-fifth and one-quarter of all federal research and development investments.

The energy R&D investment amounted to a little over ten percent of the total of all federal R&D investments during the five-year period from 1977 to 1981. From 1961 to 1977 and from 1981 to 2008, energy R&D remained a small fraction of the total federal R&D investment. Since the mid 1990s, energy R&D accounted for approximately 1 percent of the total of all federal R&D investments.

Our Investment in Energy Research and Development (R&D)

The federal government has invested about $172 billion (adjusted to 2005 dollars) in energy R&D activities during the period 1961 to 2008.

- Nuclear energy R&D, including both fission and fusion related energy R&D, represents the largest investment component: $61 billion, or slightly more than 36 percent.

- Basic energy research and science programs intended to support the more applied energy technology R&D programs accounted for $60 billion, or 34 percent of the total.

- The remaining 30 percent of the cumulative federal energy R&D investment from 1961–2008 is split: $26 billion for fossil energy R&D and $26 billion for both renewable and energy efficiency R&D.

The Kennedy, Johnson, and Nixon Energy Policy 1961–1973

During the Kennedy, Johnson, and Nixon administrations, from 1961 to 1973, federal energy R&D was dominated by nuclear energy R&D. This effort accounted for more than 70 percent of all federal energy R&D investments. Other non-nuclear energy R&D investments during this time frame were quite small. These R&D activities were scattered throughout the various federal departments and agencies and reflected the government's benign approach to energy management as a whole during this time period.

Nuclear fission energy R&D was the overwhelming focus of the federal energy R&D efforts from 1961 to 1973. The effort was focused on new reactor designs, improved uranium enrichment technologies, and on improving reactor safety, and by 1967, the technology was well understood. This understanding stemmed from the federal governments virtual monopoly on knowledge relating to fission and fusion reactions, which flowed from the nation's nuclear weapons program. In 1963 the Jersey Central Power and Light Company announced the first nuclear power plant in the United States. It was selected on purely economic grounds without the aid of the federal government. By 1967, our electric utilities had ordered seventy-five base load light-water nuclear reactors. During 1967, the Atomic Energy Commission's Light-Water Reactor Development Program was largely finished and was successful.

The success of the light-water reactors led the government to refocus the future nuclear energy R&D program. There was considerable concern during the late 1960s that the global uranium supplies were limited. As a result, the federal nuclear energy R&D effort switched their focus from the light-water reactors to the breeder reactors to ensure the continued viability of nuclear power. Federal support for fusion energy R&D comprised about 10 percent of the federal energy investment during the Kennedy, Johnson, and Nixon administrations.

Significant "energy crises" occurred during the time period from 1961 to 1973. The great blackout of 1965 occurred during the Johnson administration. The blackout caused about 25 million people in the Northeast to lose electric power. During the Nixon administration, reoccurring electrical brownouts in the Northeast were common during 1971. Again during the Nixon administration, the northern states experienced a shortage of heating oil in the winter of 1972.

In both public and private decision-making, most people acted as if ever-increasing amounts of energy at ever-lower prices would always be available. As a result, the energy policies and energy R&D activities simply did not consider future contingencies like the 1973 Arab Oil Embargo energy crisis. Even though we started to see signs of national energy inadequacies from 1961 to 1973, our leaders in Washington decided that no energy policy changes were needed by mid 1973. This is an example of the government's benign management approach to deal with our nation's energy problems.

Brad Linscott

The Ford and Carter Energy Policy 1974–1981

During the end of the Nixon administration in 1974, and through the Ford Administration to the end of the Carter administration in 1981, a fundamental reassessment of the nation's energy R&D activities occurred. The reassessment was sparked by the Arab Oil Embargo that began on October 19, 1973.

During the Ford and Carter administration, the federal investments in energy R&D rose from $2.45 billion in 1974 to $7.47 billion in 1980. The dollar amounts have been adjusted for inflation in 2005 dollars. Fossil energy R&D increased from $143 million in 1974 to $1.41 billion in 1979. The investment in nuclear energy also increased during this time period. For example, the Liquid Metal Fast Breeder Reactor program appropriation increased from $643 million in 1973 to $1.69 billion by 1977. Federal spending on renewable energy grew from $32 million in 1974 to $1.36 billion in 1979. The federal efficiency R&D program started with a modest $29 million in 1974 and expanded to $511 million in 1979.

In 1975, President Ford called for the creation of a national energy independence authority that would "assist in the construction of nuclear power plants, coal-fired power plants, oil refineries, synthetic fuels plants, and other production facilities." As a result, a rise was seen in very large energy programs. The new programs included the design, development, and building of syn-fuel (synthetic fuels) facilities. Significant federal programs were funded to deploy and demonstrate the feasibility of solar and wind energy plants, and there was a significant rapid growth in resources devoted to developing nuclear power production technologies.

The federal government invested a total of about:

- $4 billion in the Synfuels (synthetic fuels) Program from 1970–1984.

- $2 billion for the Large-Scale Solar and Wind Energy Demonstration Programs from 1978–1982.

- $16 billion for development of breeder reactors from 1968–1985.

Each of the three dollar amounts, above, are adjusted for inflation in 2005 dollars.

These budgetary increases were in direct response to the sense of urgency caused by concerns over the United States dependence on imported oil during the years 1974–1980.

The Department of Energy was established by congress and started operation in 1977, during the Carter administration.[4] The concern that congress expressed in the DOE Organization Act was that our country would face a chronic shortage of certain energy resources and that this would remain a threat to the security and welfare of the country. Congress found that the existing energy programs in 1977 were fragmented and lacking focus and coordination. It concluded that strong measures were needed, including greater emphasis on programs to increase supply and reduce demand and greater coordination of programs within a single department.

Congress began the Department of Energy with three primary expectations:

- Elevation to a department level would increase the budget priority and the budget commitment for the energy problems.

- Combining the previous separate functions under one department would improve coordination and analysis of costs and benefits.

- A single department—the DOE—would be more capable of responding to energy emergencies than would the same activities scattered throughout the government.

Brad Linscott

The Reagan, G. H. W. Bush, and Clinton Energy Policy 1981–1995

During the Reagan administration (1981–1989), a major shift in the federal government's attitude toward energy R&D occurred. A growing preference for basic energy-related science was initiated in lieu of the government's previous investments for applied energy research and demonstration facilities. The Reagan administration maintained that "only in areas where these market forces are not likely to bring about desirable new energy technologies and practices, within a reasonable amount of time, is there a potential for federal involvement."

In other words, the federal government stopped funding demonstration projects and only supported funding for basic science that was related to energy production. In addition, the government assumed that the free market would bring about the desirable new energy technologies needed to solve our energy problems. The administration expected the free market to pick up and support future energy production by using the federal energy demonstration facilities as a starting point.

Proposed Reorganization of the DOE

In December 1981, President Reagan submitted his plan to the congress for a government reorganization that included the dismantling of the Department of Energy. The elimination of the DOE would dismantle a bureaucracy but keep intact its essential functions. His plan reflected a commitment to limit the role of government in energy, based on a different view of the energy problem than that held by the congress in 1977, when the DOE was initiated. The reorganization presupposed that government planning and regulation is less necessary and desirable than before. Market forces had brought about substantial energy conservation and stimulated energy production, largely in response to the oil prices increases of 1979 and 1980. The need for a cabinet-level department to administer existing energy programs and propose new energy programs could be either reduced in size or eliminated. The government role could and, hence, should be diminished. There would be no need for a separate energy policy. Rather, energy would become a part of overall economic and trade policy.

During the first six years of the Reagan administration, federal energy R&D spending fell from $6.64 billion in 1981 to $3.15 billion in 1988.

The dollar amounts are adjusted for inflation in 2005 dollars. This amounts to more than a 50 percent decrease in energy R&D funding. During this time period, many technology programs, including the breeder reactor program, the syn-fuels program, and the large-scale solar energy demonstration, were terminated. By 1981, the DOE-funded wind turbine projects were winding down, and a few years later, most federally funded wind projects were terminated.

During the four-year George H. W. Bush administration and the first three years of the Clinton administration (1989–1995), federal energy R&D investments initially grew. Growth was predominately due to large (but short-term) increases in the federal effort to develop clean coal technologies in response to the acid rain concerns. It is estimated that about $3 billion (adjusted for inflation to 2005 dollars) was invested to demonstrate clean coal technologies. By 1993, at the beginning of the Clinton administration, appropriations for clean coal technology development programs were in rapid decline. However, support for basic energy research continued to grow. The strategy was a continuation of the Reagan policies with regard to energy R&D. By 1995, the Clinton administration's basic energy science efforts comprised nearly 50 percent of all federal investments in energy R&D. For the first time federal funding for demonstration plants and other applied technology development energy programs was the minority investment in the federal energy R&D portfolio.

Our Energy Policies in Hindsight

We have invested most of our hard-earned, energy-related tax money for nuclear energy R&D since 1961. The smallest amount of our energy R&D money has been invested in renewable energy. Now, all of a sudden, renewable energy is politically our most important energy source. Renewable energy investments started about 1979, with a very meager amount of money compared to our nuclear investment that year. It is obvious that prior efforts to eliminate our need for foreign oil were terminated during the eight-year Reagan administration and the four-year George H. W. Bush administration. The Reagan energy policy, which disregarded the need for our energy independence, continued during the eight years of the Clinton administration. This little bit of history illustrates the fact that both Democratic and Republican administrations have chosen not to diminish our reliance on imported oil and natural gas. The fact is that these administrations have allowed our country to import an ever-increasing amount of foreign oil, natural gas, and electricity than ever before. Since most of our energy-related tax dollars have been invested on nuclear energy, why is nuclear power, as an energy source, being disregarded and downplayed? Maybe our oil and gas companies had something to do with maintaining nuclear power as a minor player in our energy mix.

To summarize, during a twenty-year period from 1981 to 2001, starting with the Reagan administration to the conclusion of the Clinton administration, our nation has made little progress to attain the goal of energy independence that President Ford initiated in 1975.

One can only speculate on how history might have differed if we had pursued a more robust effort to reduce our need for foreign oil. Much of the money spent to purchase foreign oil during those twenty years could have been invested here in our country. We might have been able to reduce our financial and military activities in the Middle Eastern oil countries if our need for oil imports would have been steadily decreasing rather than on the increase. Perhaps our role in the Persian Gulf War, which started on August 2, 1990, could have been greatly diminished if we did not need to import foreign oil. In addition, our European and Middle Eastern allies might have contributed larger military forces to support the combined military efforts against terrorism. Possibly the war in Iraq could have been averted if our economic reliance on imported oil had not been a factor at the time.

Our continued need for crude oil has adversely impacted our ocean shores and ocean wildlife. Two examples of such environmental catastrophes include the loss of crude oil from the tanker Exxon-Valdez in March, 1989, and the explosion of the British Petroleum offshore drilling platform in the Gulf of Mexico in April, 2010.

The Exxon-Valdez lost 10.8 million gallons of crude oil in Prince William Sound, Alaska. The crude oil covered 1,300 miles of coastline and covered 11,000 square miles. The accidental loss of the British Petroleum platform in the Gulf of Mexico resulted in the uncontrolled escape of crude oil from the one-mile-deep ocean floor for over three months. Estimates of the amount of crude oil spilled into the ocean exceeded over 100 million gallons. A crude oil spill in the ocean, near a shore line, is considered by most to be one of the most devastating human-caused events to occur in our history.

Wind Energy

Wind Turbines in Sheldon, New York

I grew up on a farm in Bennington, New York, located about twenty miles east of Buffalo. Sheldon, New York, located about five miles south of Bennington, is where seventy-six wind turbines started producing electricity for a utility company in June, 2009. Each machine is rated at 1.5 megawatts of power. Invenergy LLC, a private company with home offices in Chicago, Illinois, financed and managed the installation of the machines located in Sheldon. Their focus was on building a positive relationship with land owners, host communities, and power purchasing customers. According to the American Wind Energy Association, Invenergy is one of the top five largest owners of wind generating facilities in our country. Invenergy purchased the 1.5-megawatt wind turbines from the General Electric Company.

Jim (a fictitious name for my friend) owns a small dairy farm in the town of Sheldon. Invenergy installed two wind turbines, each rated at 1.5 MWs, on his property. Jim receives $5,500 per year for each megawatt of rated wind power installed on his property. This remuneration is standard for all property owners having one or more 1.5-megawatt wind turbines installed on their land. So with two machines on his land, Jim receives $16,500 each year from Invenergy. In addition to Jim's payment, Invenergy pays the town of Sheldon, Wyoming County, and New York State a yearly amount of money based on a fixed value for each machine installed. The town of Sheldon receives sufficient money from Invenergy so that the land owners residing in Sheldon no longer have to pay property tax on their land. Invenergy manages the operation and maintenance for each of the wind turbines located in Sheldon. In the event that the machines can no longer be maintained, Invenergy posted a bond with the town of Sheldon. The bond provides sufficient money to remove each machine in the event that removal is necessary. Jim is extremely satisfied with his agreement with Invenergy and their agreements with the town of Sheldon. He is certainly a strong advocate for wind power.

Now, my friend Jim is a happy camper. But how do we taxpayers fit into this picture? There are several congressional acts benefiting private investors for their purchase and installation of large wind turbines:

- An income tax credit of 2.1 cents/kilowatt-hours is allowed for the production of electricity from utility-scale wind turbines. The renewable production tax credit incentive was created under the Energy Policy Act of 1992. Congress acted to extend the provision to December 31, 2012. Additionally, wind project developers can choose to receive a 30 percent investment tax credit in place of the production tax credit. The benefit is allowed with the provision that the facility must be placed in service until the end of 2010. The provision is also allowed for facilities placed in service before 2013 if construction begins

Brad Linscott

before the end of 2010. This is called the "Production Tax Credit." This allows a private business a credit that applies to electricity generated from wind turbines for sale at "wholesale" to a utility company that sells electricity to customers at "retail."

- Under present law a federal-level investment tax credit is available to help consumers purchase small wind turbines for home, farm, or business use. Owners of small wind systems with 100 kilowatts of capacity or less can receive a credit for 30 percent of the total installed cost of the system. The value of the credit is now uncapped as a result of the law enacted in 2009.

The production tax credit imposes a cost to taxpayers in the form of forgone federal tax revenue.[5] Renewable energy claims for the tax credit have increased significantly over time from just $4.0 million in 1995 (adjusted to 2005 $) to over $210 million in 2004. Wind projects accounted for over 90 percent of the dollar value of the renewable energy tax credit claims through 2004. Tax credit claims are estimated at $850 million in 2007 and increasing to maximum of $1.4 billion in 2009. Since the tax credits inception, from 1995 through 2009, the loss of treasury revenue resulting from renewable energy claims is estimated at $6.6 billion.

Renewable electricity is not the only energy source to receive federal fiscal incentives. Tallying federal subsidies is tricky business, but one estimate sets average federal fiscal subsidies for 2006 at $75 billion. Of the $75 billion in subsidies, fossil fuels received $49 billion, nuclear energy received $9 billion, and ethanol received $6 billion.

Wind Turbine Operation

Wind turbines are only able to produce electrical energy when winds are available and only when wind speeds are within a narrow range.[6] If wind velocities are low, 5 to10 mph or less, the machines will not start. When wind velocities are too high, usually above 40 mph, the machine are programmed to shut down or operate at reduced power. Winds tend to be more available at night and during the colder months. Electric utilities experience their highest demand during the summer months and particularly during the afternoon. As a result, the wind turbines cannot be relied on to provide our electric utilities with energy during their peak summer-afternoon demands. Wind farms have to be located in remote areas of our country where winds are available. The remote areas where wind turbines are placed usually require new electric power transmission lines to connect the wind turbines to existing utility transmission networks.

On average, the large wind turbines are able to operate only about 35 percent of the year, or 128 days of each year. At best, wind turbines can only be relied on to save fossil fuels. The majority of our electric power plants use natural gas or coal as their energy source. As a result, wind turbines used to generate electricity will not reduce our dependence on imported oil.

Wind-turbine-generating capacity, in our country, totaled 11,329 megawatts in 2006 and increased to 16,515 megawatts in 2007. Texas continues to lead the nation's wind power development with 1,752 megawatts of new wind capacity placed in service in 2007. Texas has the leading share—27 percent of the nation's wind turbine capacity in operation at the end of 2007. Sixty-two percent of the total wind-generating capacity is located in Texas, California, Iowa, Washington, and Minnesota.

Costs for Wind Turbine Electricity

The National Renewable Energy Laboratory (NREL) in Golden, Colorado, has developed a reliable tool for estimating the cost of wind-generated electricity.[7] Their model is capable of projecting the cost of electricity for both land-based and offshore wind turbines. The models are intended to provide reliable cost projections for wind-generated electricity based on wind turbine size. The models are not intended to predict the "price" of a wind turbine as delivered by a manufacturer. Wind turbine prices may be impacted by various economic conditions and other volatile market factors. As a result, these factors were not included in their models for estimating the "cost" of wind-generated electricity. Cost estimates are projected based on the turbine's rated power (MW), the rotor diameter, the height of the cylindrical tower, and other typical components that comprise today's modern wind turbine.

The NREL report provides two cost-breakdown estimates as examples for two wind turbines of different size. The first example tabulates the cost estimate and cost of energy for a land based 1.5-megawatt wind turbine in 2005 dollars. The second example presents the cost estimate and cost of energy for an offshore (shallow water) 3.0-megawatt wind turbine in 2005 dollars. The costs, presented in tabular form, are for each of the machines' main components, including the blades, tower, and a number of other items. Itemized costs are also presented for the foundation, transportation, assembly, and other activities related to installation of the wind turbine at the site. The total costs are summarized to arrive at an estimate for the installed cost and the installed cost of energy. The tabular cost information presented for the 1.5-megawatt machine and the 3.0 -megawatt machine can be used in a comparative way.

Cost for the 1.5-megawatt Wind Turbine on Land

The initial capital cost from the manufacturer for the land-based 1.5-megawatt machine is estimated at $1.0 million. On-site activities include the foundation, transportation, roads, civil work, assembly, installation, electrical interface connections, engineering, and permits. The estimated cost for on-site activities amounts to $367 thousand. The machine cost of $1.0 million added to the $367 thousand cost for on-site activities amounts to an estimated capital cost of

$1.4 million for the 1.5-megawatt wind turbine. The land-based on-site activities amount to 26 percent of the total installed cost for the 1.5-megawatt machine.

Cost for the 3.0-megawatt Wind Turbine in the Ocean

Making the same comparison for the 3.0-megawatt wind turbine, the initial capital cost from the manufacturer is estimated at $2.7 million. The capital cost includes $321 thousand for protecting the machine from the adverse effects of its location in shallow seawater. The estimated cost for the on-site activities in seawater is estimated at $3.7 million. The machine cost of $2.7 million added to the $3.7 million cost for on-site activities amounts to an estimated capital cost of $6.4 million for the 3.0-megawatt machine. The cost for the on-site installation in seawater is $1.0 million more than the cost just to produce the 3.0-megawatt machine. The shallow-ocean-water-site activities amount to 58 percent of the total estimated installed cost for the 3.0-megawatt wind turbine.

To summarize, the installation cost amounts to 26 percent of the total installed cost for the 1.5-megawatt machine sited on land and 58 percent for the 3.0-megawatt machine sited in the ocean. The comparison shows that the installation cost for a wind turbine sited in shallow ocean waters will be considerably higher than for a wind turbine sited on land. The US Energy Information Administration anticipates offshore construction will cost more than onshore construction and assumes that onshore locations are abundantly viable and economically attractive.

Costs for Wind Turbine Operation and Maintenance (O&M)

Another factor considered by NREL is their estimate for the operation and maintenance (O&M) costs, on a yearly basis, for each wind turbine. Because the machines are of different size, the ratio of O&M cost divided by the machine rating in megawatts provides a more realistic comparison of O&M costs for each machine. It is expected, for example, that replacement parts for the 3.0-megawatt machine will be more expensive to replace than similar parts for the 1.5-megawatt machine because of size and weight differences. As an example, for the 3.0-megawatt wind turbine, each blade is longer and heavier than a similar blade for the 1.5-megawatt machine. Each of the three blades mounted on each machine is similar in appearance to the wings on a glider aircraft. Therefore, the estimated O&M cost for the larger machine can justifiably be higher. The estimated O&M cost for the 1.5-megawatt machine is $30,000 per year and $215,000 per year for the 3.0-megawatt machine. The ratio of O&M dollars per year divided by the rated power for the 3.0-megawatt wind turbine is $71,600 per megawatt and $20,000 per megawatt for the 1.5-megawatt wind turbine. As expected, the cost to operate and maintain a wind turbine in seawater can be at least three times higher than a wind turbine sited on land. Some of the hazards associated with offshore operation include human safety during regularly scheduled maintenance and repair activities, wind turbine icing, environmental impact on lake and ocean beds, and the possible collision of boats and ships with wind turbine foundations.

To summarize, the cost of energy produced by the 3.0-megawatt machine sited in seawater is twice as much as the 1.5-megawatt machine on land. The difference in the cost is primarily due to the increased effort needed to install the large wind turbine on the ocean floor. Keep in mind that these are *cost* numbers that will be lower than the *price* consumers will pay. These cost estimates do not include the increased amount that a utility or a private investor must charge in order to show a profit.

A large wind turbine will produce electricity at a lower cost per kilowatt than a smaller machine of similar design when both are located at the same wind site. Part of the reason for this is that the larger machine, having longer blades installed on a higher tower, can capture winds at a higher altitude. Because wind velocities tend to increase with altitude, the larger machine is able to collect more wind energy and produce more electric power. But the NERL model shows just the reverse results. The cost per kilowatt of electricity produced by the 3.0-megawatt machine installed in the ocean is twice as much as the 1.5 -megawatt wind turbine installed on land.

What is the Capacity Factor?

The capacity factor for an electric power plant is determined by a ratio, expressed as a percentage, of the actual number of days of operation divided by the number of days in one year. The Northwest Power Planning Council uses 33–34 percent for a capacity factor when estimating the output of wind farms in the states of Washington and Oregon. Because of the unpredictability of the wind and other factors, these wind turbines, on average, will only produce energy about two and a half days out of seven. The much-advertised, smaller, and older wind turbines installed in California during the 1980s are estimated to have a capacity factor of between 5 and 20 percent. In other words, these machines are producing electrical energy, on average, only between a third of one day and one and a half days each week.

The 3.0-megawatt machine is estimated to have a capacity factor of 38.1 percent because of its offshore location. The cost of energy in dollars per kilowatt for the 3.0-megawatt machine is estimated at $0.095. The 1.5-megawatt wind turbine is estimated to have a lower capacity factor than the 3.0-megawatt machine because the winds are less steady on land than over the ocean. The 1.5-megawatt wind turbine on land is estimated to have a capacity factor of 32.8 percent, and the cost of energy is estimated in dollars per kilowatt at $0.0476. The capacity factor for wind turbines is much lower than our nuclear plants. Our nuclear power plants that operate today have a capacity factor of about 95 percent. In other words, they are able to produce electrical energy 347 days, 24 hours per day, out of 365 days each year.

Should We Pay for More Wind Turbines?

During 2008, T. Boone Pickens was advertising the virtues of using wind turbines and natural gas to solve our energy problems.[8] He runs the Dallas-based energy investment fund BP Capital. He spent about $60 million during his advertising campaign in an effort to reduce the nation's reliance on foreign oil.

Pickens's company, Mesa Power, ordered 687 wind turbines (an investment of about $2 billion) from the General Electric Company in 2008. In addition, he has leases on about 200,000 acres of land in the Texas Panhandle. He had planned to install the world's largest wind farm on the Texas property where he holds the leases. Pickens uncovered a problem in getting power from the proposed wind turbine farm in the Panhandle to an existing electric utility network. As a result, his plans for the large wind farm in the Texas Panhandle have been scrapped. He is looking at potential wind sites in the Midwest and in Canada. His TV advertisements in 2009 suggest that natural gas is now the answer to meet our energy needs. It is further suggested that we need to drill more oil and gas wells to meet a potential future demand for gas.

There is a perception that renewable energy will free us from our dependence on the need for imported oil. The idea that wind turbines generating electricity will diminish our need for foreign oil is incorrect. We don't use foreign oil as a fuel for our large electric power plants. Our fossil-fuel-fired electric plants use primarily our own coal and natural gas. A small fraction of our electric plants use gasoline or diesel fuels to produce electric energy. The large electric plants, using either natural gas or coal, can reduce their expenditure of fuel when wind turbines are generating power on line. Wind turbines can only reduce, by a small amount, the natural gas and coal that our electric power plants consume. As an example, 40,000 wind turbines, rated at 1.5 megawatts each, would supply only 5 percent of the energy needed if they were used to augment our electric energy production in 2007.[9]

Solar Energy

As long as the sun is shining, solar energy is available. The energy from the sun striking the earth for just forty minutes is equivalent to the global energy consumption for one year. It causes no pollution, there are no gaseous emissions, and it's an environmentally friendly source of energy. Solar energy is prevalent all over the surface of our earth and millions of miles above the surface of our world.

Wind energy sites, on the other hand, must be selected, for reasons of economy, where average wind speeds are the highest. Solar energy can be gathered above the surface of the earth, unlike fossil fuels, which require excavating the surface of the earth or drilling below the earth's surface or fracturing underground layers of rock.

The two primary methods for extracting energy from our sun are:

- The photovoltaic cell
- The solar energy concentrator

Photovoltaic Cells

The modern solar cell was patented in 1946. The solar cell was discovered during research efforts that led to the transistor. Solar cells
are electrically connected and encapsulated as a unit called a *module*.
The modules are covered with a thin sheet of glass to protect the
solar cells from the elements that result from adverse weather conditions. Modules are connected together in groups to form an array.
The solar cells, depending on how they are connected electrically, are
able to produce high voltage or high current. In the past few years
the cost of photovoltaic cells and modules has dropped significantly.

Various types of solar voltaic cells exist. The least expensive of all
of the cells being considered are solar voltaic cells called *thin film cells*
made of cadmium telluride. To provide electricity for six cents per
kWh, the thin cells would have to convert solar energy to electricity
with an estimated efficiency of 14 percent. Thin cells have an efficiency of only 10 percent. In order to provide electricity at six cents
per kWh, the price to fabricate, assemble, and install an array of thin
cells has to about $1.20 per watt of power produced.

What was the price in 2009? About $4.00 per watt. Because solar
cells have an efficiency of about 10 percent, huge tracts of land are
needed to produce the large amounts of electrical energy that our
country uses every day. As the efficiency of solar cells increase, the
area of land needed will diminish. Another challenging factor is that
large tracts of land are required because solar cells produce energy
less than 50 percent of the time. Solar advocates suggest the need
for as many as 30,000 square miles of solar arrays. If 30,000 square
miles of solar arrays were placed in the northeastern portion of our
country, they would completely cover the states of Massachusetts,
New Hampshire, and Vermont.

Solar arrays produce direct current (DC) electricity. In order
for the utilities to use solar energy, the DC has to be converted to
alternating current (AC). The majority of future large solar energy

Brad Linscott

farms requiring huge tracts of land have to be located in the south-west corner of the country, where sunshine is most prevalent. The large majority of our electrical energy is used in the northeastern area, the northern midwestern area, and the central midwestern area. Some solar energy advocates propose that we build a new DC-high-voltage electric network of utility lines to carry the energy across the nation. Our electric energy transmission lines are designed and built to transmit high-voltage alternating current. They are not capable to efficiently carry direct current high-voltage energy.

Covering 70 million homes in the United States with solar panels would produce only 1/10 of the electric power consumed in the year 2000.[10] That is why land will be needed and dedicated for solar farms. Using large tracts of land to collect solar energy will raise environmental concerns, such as destroying the local habitat where the solar farms are sited. As mentioned before, the photovoltaic solar cells only produce energy when the sun is shining. To provide electricity during the nighttime hours and cloudy days, methods for storing electrical energy are being investigated. One of the most difficult tasks facing the collection of solar energy is that of determining the best way to store electric energy.

Small Photovoltaic Power Systems

The DOE and NASA has helped to foster new private companies that can profitably design, build, and sell solar-electric devices. One example of a new company is the GreenField Solar Corp., located in Cleveland, Ohio. The company is planning to market a 1.5-kilowatt solar-electric power plant.

For the sunshine states, solar voltaic cells help to reduce homeowners' electric bills. When the price for fossil fuels is high, solar electric energy makes economic sense. Homeowners can get a one-time federal tax credit of as much as $2,000 on an installation, but the credit was scheduled to expire in 2009. A proposed extension during 2010 would extend the credit's cap to as much as $4,000 and open it up to individuals who make enough money to qualify for the alternative minimum tax. Some states, such as California, New York, and Connecticut, have their own subsidies.

Storage for Electrical Energy

In order to store energy, electricity must be converted to some other form of potential energy. One possibility that is being considered is to use the electricity to pump water uphill to a reservoir. The water could be later released through hydroelectric generators to produce electricity when needed. Another possibility is to use the electrical energy produced by solar cells to generate hydrogen gas. The hydrogen can then be delivered via pipelines to the locations where it could be used to fuel electric power plants.

To improve the viability of the wind and solar farms, our tax money is being used to research and develop ways of pumping air into vacant underground caverns, abandoned mines, aquifers, and depleted natural gas wells. The electricity produced by the wind turbines and the solar arrays would be used to pump air into geological underground cavities at pressures up to 1,100 pounds per square inch. It is estimated that facilities acquired for and prepared to accommodate compressed air storage would add at least three to four cents per kWh to the cost for photovoltaic and wind energy generation.

Pressurized air stored in underground cavities would be released on demand through a proposed newly built infrastructure of pipelines needed to deliver the high-pressured air to gas-fired electric plants located all across the United States. The cost for design and construction of a compressed-air pipeline infrastructure would significantly add to the already high price for solar and wind energy. Gas-fired electric plants can use a mixture of compressed air and natural gas for combustion. By mixing compressed air with natural gas, the turbines can produce the same amount of power with only

40 percent of the natural gas that normally would require 100 percent natural gas. Each gas-fired power plant using the gas and air mixture would require modification to accept the new fuel mixture.

Does your common sense tell you that we should store compressed air underground and build an extensive pipeline across our country to transport compressed air to our gas-fired power plants? The cost and legal aspects of gaining the right-of-way from individual property owners will pose huge problems. We are currently installing new pipelines across the country to transport natural gas from people's backyards to natural gas consumers. Do we really want more pipelines crisscrossing our country to transport compressed air and natural gas?

Solar Concentrators

Concentrating solar energy technologies use mirrors to reflect and concentrate sunlight onto receivers that collect the solar energy and convert it to thermal energy. For example, solar energy can be used to convert water into a form of thermal energy commonly known as *steam*. Superheated steam is used to drive a steam turbine connected to an electric generator. The thermal energy may also be used to power a heat engine connected to an electric generator.

There are three main technology systems that fall under the category of solar concentrators. They are:

- The Linear Concentrator
- The Dish/Engine System
- The Power Tower System.

Linear Concentrators

Linear-concentrating collector fields consist of a large number of highly reflective mirrors, called *collectors*, oriented in parallel rows.

They are typically aligned in a north-south orientation to maximize both annual and summertime energy collection. With a single-axis sun-tracking system, this configuration enables each collector to track the sun from east to west during the day.

The shape of the metal mirrors is rectangular in plan form, having two parallel sides much longer than the other two sides. The collectors are mirrors made of a metal such as an aluminum alloy. The collectors capture the sun's energy by reflecting the sunlight onto long pipes called *receiver tubes*. The receiver tube contains a fluid that is heated by the sunlight. The hot fluid is piped into a heat exchanger. A typical heat exchanger is designed with two compartments separated by a common wall. The hot fluid flows into and out of the first compartment of the heat exchanger. Water flows into and out of the second compartment of the heat exchanger. Inside the heat exchanger, the hot fluid transfers its heat through the common wall from the hot fluid compartment into the cold fluid compartment. Sufficient heat is transferred through the common wall to turn the cold water into superheated steam. The steam is used to drive a turbine and generator to produce electricity. Alternately, steam can be produced directly in the receiver tubes. This concept eliminates the need for heat exchangers.

The predominant concentrator system that is operating in our country is the linear concentrator having a parabolic trough shape. The parabola has a focal point at which all of the solar energy is centered. The parabolic trough design provides a large number of focal points. They are so numerous and close together that the points form a focal line. The receiver tube is positioned along the focal line of each parabolic-shaped collector. The tube is fixed to the collector structure. The cool fluid is heated by solar energy as the fluid flows into and along the length of the receiver tube. As the fluid flows out of the receiver tube it is transported by pipeline to where it is used. For the case of a water/steam receiver, steam is sent directly to the turbine.

Trough designs can incorporate thermal storage. In such systems, the collector field is oversized to provide heat to a storage system during the day. The storage system can be used in the evening or during cloudy weather to generate additional steam to the turbine- generator. Parabolic trough power plants can also be designed as hybrids, meaning that they use fossil fuel to supplement the solar output during periods of low solar radiation. In such a design, a natural-gas-fired heater or gas-steam boiler/re-heater is used. In the future, troughs may be integrated with existing or new combined-cycle natural-gas and coal-fired plants.

A large individual trough system, called "Nevada Solar One," generates 64 megawatts of electricity. The plant, located in Boulder City, Nevada, uses 760 parabolic concentrators consisting of 182,000 mirrors. The mirrors concentrate the solar energy onto more than 18,000 receiver tubes. Nevada Solar One covers about 400 acres of land and started producing electricity in June, 2007.

Dish/Engine Systems

The dish/engine system is a concentrating solar power technology that produces relatively small amounts of electricity compared to the linear concentrator and the power tower systems. The dish/engine power system is typically rated between 3 to 25 kilowatts of electricity. The solar concentrator is dish-shaped. The shape can be visualized as being formed by a parabola revolved about its axis of symmetry. A thermal receiver is located at the focal point of the parabolic dish where the solar energy is concentrated. The dish is mounted on a structure that tracks the sun during the daylight hours for the purpose of maximizing the solar efficiency of the system. If the sun is not aligned with, or not parallel with, the axis of symmetry of the dish, the location of the focal point will shift away from the location of the receiver. In this case, the receiver will not receive the maximum amount of solar energy reflected by the dish.

A power conversion unit includes two subsystems: the *thermal receiver* and the *engine/generator*. The thermal receiver is the interface between the dish and the engine/generator. It absorbs the concentrated beams of solar energy, converts them to heat, and transfers the heat to the engine/generator. A thermal receiver can be a bank of tubes filled with a working fluid. The fluid is usually hydrogen or helium and is used to absorb the heat energy delivered from the dish to the receiver. The heated fluid flows from the receiver to a heat engine. The fluid transfers its thermal energy to the heat engine and flows back to the receiver to be reheated.

The engine/generator subsystem takes the heat from the thermal receiver and uses it to produce electricity. The most common type of heat engine used in the dish/engine power plant is called the Stirling engine. A Stirling thermodynamic cycle engine uses the heated fluid to move pistons and create mechanical power. It is called a heat engine because it can produce energy without the need to burn fuel. Solar energy can be used to power the engine, and the mechanical work produced by the engine's drive shaft turns a generator that produces electricity.

Watching a laboratory model of the Stirling engine operate is quite astonishing. The models are typically made of clear plastic so that the internal parts of the engine—including the piston, connecting rod, and crankshaft—can be observed during operation. A portion of the engine is made of metal where heat is applied. As the heat is applied, the engine begins to quietly start running. No exhaust emissions are produced by the solar-powered Stirling engine.

Power Tower Systems

The power tower system uses a large number of flat, sun-tracking mirrors to focus sunlight onto a receiver located at the top of a tower. Each large mirror is called a *heliostat*. A heat-transfer fluid is circulated in the receiver. The fluid is heated by solar energy reflected

Brad Linscott

from the mirrors and pointed toward the receiver. The heated fluid in the receiver is used to generate steam, which in turn is used by a conventional turbine generator to produce electricity. Newer power tower systems are using molten salt nitrates as the working fluid because of its superior heat-transfer and energy storage capabilities. Individual commercial plants can be sized to produce up to 200 megawatts of electricity.

During its operation from 1982 to 1988, the DOE-funded, 10-megawatt Solar One plant located near Barstow, California, demonstrated the viability of power towers. The plant concentrated reflected solar energy from 2,000 mirrors onto a heat exchanger called a *receiver unit* mounted at the top a 300-foot-high concrete tower.[11] The receiver was designed to absorb the solar energy and convert water into steam. The steam was used to power a steam-turbine electric generator. The Solar One plant was modified and called Solar Two. Solar Two demonstrated the advantages of molten salt for its heat transfer and thermal storage capabilities. Solar Two successfully demonstrated efficient collection of solar energy and its conversion to electric power. It also demonstrated the ability to routinely produce electricity during cloudy weather and at night. In 1995, Southern California Edison, the operator, shut the plant down because it wasn't commercially viable.

While the NASA Wind Energy project office was addressing the issues related to wind turbine operations, a story emerged in the office about an early experience with the DOE Solar One project located in California. Apparently the 2,000 mirrors, a key component of the Solar One power tower, accumulated dew-like moisture overnight that coated the reflective surfaces. Close to dawn, a brisk wind blew up before the moisture evaporated. The wind raised sand from the ground. The sand carried by the wind was deposited on the wet reflective side of most of the 2,000 mirrors. The story went on to describe the efforts of the maintenance crew to remove sand from each one of the 2,000 mirrors. The crew worked with mirror-cleaning solutions and towels for many hours to carefully remove the residue, while being careful not to harm the reflective quality of each mirror. I'm not sure this "grapevine" story is completely factual, but the wind office experienced some solace in knowing that renewable energy issues were not isolated to just wind turbines.

Thermal Storage

One of the challenges facing the widespread use of solar energy is the reduced or curtailed energy production when the sun sets or is blocked by clouds. The power plant must be initially designed to collect an "excess" amount of solar energy during the daytime. In other words, the rated power of the solar plant is designed to be higher than the average power it can generate during a twenty-four-hour period. The excess thermal energy produced during the daytime is stored for use during the night time or a cloudy period.

The DOE is developing the technology that will lead to practical methods to store thermal energy. The stored thermal energy can be used to produce electricity during the nighttime hours and during cloudy days. The DOE is developing three technologies for energy storage:

- The Two-Tank Direct System
- The Two-Tank Indirect System
- The Single-Tank Thermocline System.

Two-Tank Direct System

With this system, thermal energy is stored in the same fluid used to collect the thermal energy. The fluid is stored in two tanks, one at high temperature and the other at a lower temperature. Fluid from the low-temperature tank flows through the receiver tubes of the trough-type solar collector or a receiver used by either the parabolic dish collector or the power tower. Solar energy heats the fluid to a high temperature as it flows through the receiver. The heated fluid flows through a pipe to the high-temperature tank for storage. With another pipe, fluid is drawn from the high-temperature tank and directed into a heat exchanger. As the high-temperature fluid flows through one side of the heat exchanger, water is circulated on the other side of the heat exchanger. The water is turned into steam by the heat transferred from the high-temperature fluid. The steam is used to drive a turbine connected to an electric generator. The high-temperature fluid is cooled as it flows through the heat exchanger because it is transferring heat to the water. The fluid exits the heat exchanger, now at a lower temperature, and returns to the low-temperature tank. Two-tank direct storage was used in early parabolic trough power plants and at the Solar Two power tower in California. The trough plants used mineral oil as the heat transfer and storage fluid. Solar Two used molten salt as the heat transfer and storage fluid.

Two-Tank Indirect System

The two-tank indirect system functions the same as the two-tank direct system, except two different fluids are used instead of one type of fluid. The first fluid is called the *heat-transfer fluid*, and the second fluid is called the *storage fluid*. The two-tank indirect system requires two heat exchangers instead of just one. This system is used in power plants where the heat-transfer fluid is too expensive or not suited for use as the storage fluid. The storage fluid from the low-temperature tank flows through the number-two heat exchanger, where it is heated by the high-temperature heat-transfer fluid. The hot storage fluid flows back to the high-temperature storage tank. The heat-transfer fluid exits the number-two heat exchanger at low temperature and returns to the receiver tubes of the trough-type solar collector or to a receiver, where it is heated back to a high temperature.

Storage fluid from the high-temperature tank flows through the number-one heat exchanger. The hot storage fluid transfers its heat to the water side of the number-one heat exchanger. As the water warms and turns into steam, the steam is used to drive a turbine-electric generator in the same manner as the two-tank direct system. The indirect system requires one additional heat exchanger, which adds to the cost. This system has been proposed for several parabolic collector types of power plants that will use organic oil as the heat transfer fluid and molten salt as the storage fluid.

Single-Tank Thermocline System

This system stores thermal energy in a solid medium, most commonly silica sand placed in a single tank. During operation of the power plant, a portion or the top layer of the sand inside the tank is at high temperature and a lower layer of sand at the bottom of the tank is at a low temperature. The hot sand at the top and the cold sand at the bottom are separated by a middle layer of sand that varies in temperature. The upper portion of the middle layer of sand is at the high temperature. The lower part of the middle sand layer is at the low temperature. As a result, the temperature of the middle sand layer varies from hot at the top to cold at the bottom. This variation of the sand temperature is called a *thermocline*. The temperature of the sand halfway between the top and bottom of the middle layer is likely to have a temperature halfway between the hot temperature at the top and the cold sand temperature at the bottom of the tank.

During the energy storage process, the high-temperature heat-transfer fluid flows into the top of the tank, passes through each of the three layers of sand, and exits out the bottom of the tank at a low temperature. By continuing the flow of hot fluid into the tank, the volume of sand at high temperature begins to increase. The volume of sand at the cooler temperatures begins to decrease as the volume of hot sand increases. The result of this process adds thermal energy to the total volume of sand inside the tank. Once the sand has absorbed sufficient thermal energy, the energy can be removed from the tank by reversing the flow of the heat-transfer fluid. The cold heat-transfer fluid can be piped into the bottom of the tank and forced to flow to the top of the tank, where the heated fluid is used generate steam to operate the turbine electric generator.

In using sand as a storage medium and only one storage tank, the cost for this system is considerably less compared to the two-tank systems. The single-tank thermocline system was demonstrated at the Solar One power tower, where mineral oil, instead of sand, was used as the storage fluid and steam was used as the heat-transfer fluid.

Hydropower, Ocean Tides, and Wave-Action Hydropower

Examples of large hydroelectric power plants located in our country include the Hoover Dam, The Grand Coulee Dam, and the Robert Moses Niagara Power Plant. The Grand Coulee Dam is the largest producer of hydropower in our country. It is rated at 6,800 megawatts. Grand Coulee is located in the state of Washington on the Columbia River. The Robert Moses Plant is located in Lewiston, New York, near Niagara Falls, New York. The Robert Moses plant has a rated generating capacity of 2,515 megawatts. The Hoover Dam near Las Vegas, Nevada, is located on the Colorado River separating the states of Nevada and Arizona. Hoover Dam has an installed capacity of 2,080 megawatts.

Conventional hydroelectric power continues to decline as a share of the total net generation. In 2007, conventional hydroelectric-generating capacity accounted for 6 percent of the total net generation compared to 10.2 percent in 1997. Renewable energy sources, excluding conventional hydroelectric generation, contributed 2.5 percent of the total net electric generation in 2007.

A decline in conventional hydroelectric generation is consistent with the drought conditions, which according to the National Climatic Data Center, prevailed over the West and Southwest for much of 2007. Evaporation caused by above-normal summer temperatures exacerbated drought conditions along with below-average precipitation in these two areas.

Ocean Tides

Some of the oldest ocean energy technologies use tidal power.[12] All coastal areas consistently experience two high and two low tides over a period of slightly more than twenty-four hours. For those tidal differences to be harnessed into electricity, the difference between high and low tides must be more than sixteen feet. There are only about forty sites on the earth with tidal ranges of this magnitude. Currently there are no tidal plants in the United States. However, conditions are good for tidal-power generation in both the Pacific Northwest and the Atlantic Northeast regions of our country.

Tidal-power technologies can be divided into three categories: the hydroelectric dam, a tidal fence, or a tidal turbine.

Hydroelectric Dams

A dam similar to the Hoover Dam but smaller in size is typically used to convert tidal energy into electricity by flowing water through turbines. The turbines connected to electric generators are located at the bottom of the dam. Dam gates and a series of water turbines are installed at the bottom of the dam, along the length of the dam. At high tide the gates are closed. As the tide recedes and when the tides produce an adequate difference in the level of the water on opposite sides of the dam, the gates are opened. The water then flows through the turbines. The gates are left open until high tide is reached. The gates are then closed and the cycle repeats itself.

Tidal Fence

A tidal fence operates like a turnstile. The tidal fence is designed to reach across channels between small islands or across straits between the mainland and an island. Tidal currents, typical of coastal waters, force the tidal fence to rotate similar to the turnstile. The rotational energy is used to drive an electric generator. Some of the ocean currents reach velocities of between five and nine miles per hour. Seawater has a much higher density than air. As a result, ocean currents carry significantly more energy than the air currents used by wind turbines.

Tidal Turbines

Tidal turbines look much like a wind turbine but operate under water. They are arrayed underwater in rows. The turbines function best where the velocity of coastal water currents run between 4 and 5.5 miles per hour. With water currents in this speed range, a tidal turbine with a blade diameter of fifty feet can generate as much energy as a wind turbine having a blade diameter of two hundred feet. Ideal locations for tidal turbine farms are close to shore in water depths between sixty-five and ninety-eight feet deep.

There are currently three tidal plants that are operating in the world today.[13] One is located in France, one in Russia, and one in Nova Scotia. All three tidal plants use dams to hold the water before releasing the water through a water-turbine generator. This design is similar to conventional hydroelectric power plants. Nova Scotia's Tidal-Generating Station has been operating since 1984. It uses the Bay of Fundy tides to produce twenty megawatts of energy.

Similar to all of the energy alternatives tidal power plants have their own set of environmental and economic challenges. Tidal power plants that dam estuaries can impede sea-life migration. The buildup of silt behind these facilities can impact local ecosystems. Like tidal dams, tidal fences may disturb sea-life migration. Newly developed tidal turbines may prove ultimately to be the least intrusive to the water ecosystem because these turbines do not block migratory paths

The Nova Scotia government together with the Nova Scotia Department of Energy has moved in a direction to build and test three or more candidate tidal energy facilities. The new in-stream tidal technology, when fully developed, has the potential to generate three hundred megawatts of "green," emission-free energy from two locations in the Bay of Fundy. The Nova Scotia government is sponsoring an environmental assessment to identify the potential impact of the proposed facilities on marine life and fisheries. The results of the assessment will help decide how best to develop the tidal resource. The government plans to develop a streamlined policy framework to assist future developers of this resource. The government is also inviting developers from around the world to demonstrate in-stream tidal devices as part of a common demonstration facility located in the Minas Channel area of the Bay of Fundy.

Marine energy experts have identified what they believe to be an ideal site for the Nova Scotia's first demonstration turbines. The site is in the Minas Passage area of the Bay of Fundy near Black Rock. Some of the reasons for selecting this site include: water depths of up to 148 feet at low tide, a sediment-free bedrock seafloor, and straight-flowing water currents with speed up to twenty-two miles per hour during tidal ebb and flow.

In the United States, Natural Currents New England plans to develop six tidal/ocean sites in the Northeast and two in the Northwest. In the Northeast, one of the sites selected will be located on the Cape Cod Canal. If these plans culminate, a 30-megawatt system will be constructed that is expected to generate clean, renewable, electric energy. The project received a federal permit to start construction.

Wave Power

Waves are the result of the wind blowing across the ocean's surface and are also caused by the ocean tides. Waves are constantly forming and crashing across the surface of the oceans and provide a continual source for extracting energy. The west and northeast coasts of the United States, the west coast of Europe, and the coasts of Japan, New Zealand, and South Africa are prime locations for the harvest of wave energy.

Wave energy can be harnessed on shore or off shore. On shore, the waves can be focused into a narrow, tapered channel, which increases the power and size of the waves. These new, "enhanced" waves are then used to power water driven turbines. For offshore wave power, the bobbing motion of waves can be used to mechanically extract energy.

Several designs are proposed to capture the energy from waves.[14] Some of the more promising designs are undergoing demonstration testing that could have commercial application. Offshore systems are situated in deep water, typically of more than 130 feet. Wave energy devices are intended to be installed at or near the surface. These devices convert the energy of the waves into other energy forms, usually electricity. Two of the designs have been the target of development. The *oscillating water column* and the *point absorber* are designed and being tested to capture wave energy.

Oscillating Water Column

The *oscillating water column* is a cylindrical device initially filled with air. The cylinder is open at the bottom and closed at the top. The bottom of the cylinder is secured by a cable or chain to the ocean floor. A two-way air valve is attached to the top of the cylinder located above the water surface and allows air to flow in and out of the cylinder. The upward wave action causes the air column inside the cylinder to be compressed. The compressed air is used to drive an air turbine that in turn drives an electric generator. The two-way valve at the top above water allows the cylinder to refill itself with air during each downward-wave cycle.

Point Absorber

A second device is called a *point absorber*. The point absorber consists of two basic parts—a long cylindrical container and a long cylindrical rod located inside of the container. The length of the cylinder and rod are oriented perpendicular to the ocean floor. The upper portion of the cylindrical container is designed to float above the surface of a wave, while the lower portion remains below the water's surface. The interior of the cylinder contains permanent magnets. The permanent magnets are secured to the inner wall of the containment cylinder. One end of the cylindrical rod extends through a water-tight seal at the bottom of the cylinder and is secured to the ocean floor by a cable or chain. The upper portion of the cylindrical rod inside of the cylinder is designed like the armature of an electric generator. As the vertical cylinder bobs up and down due to wave action, the armature remains stationary because it is fixed to the ocean floor. The resulting linear motion of the magnetic field surrounding the fixed armature produces electricity.

The classical direct-current electric generator produces electricity by rotating the wire wound armature within a stationary magnetic field. A linear generator usually operates by moving the armature in a straight line within a stationary magnetic field. The point absorber functions in the opposite manner compared to the linear generator. The armature attached to the long rod remains stationary because one end of the rod is fixed to the ocean floor. The magnetic field is now the moving part of the device because the magnets are fixed to the oscillating cylinder.

Some of the environmental considerations for the development of wave energy are:

Brad Linscott

- Positive or negative impacts on marine habitat depend on the design of the energy device. Submerged surfaces of the device may attract marine life that could adversely affect its operating efficiency.

- Toxic releases from leaks or accidental spills of liquids used as part of a particular device must be considered. For example, some designs similar to the oscillating water column use hydraulic oil instead of air as the working fluid.

- Visual and noise impact on marine habitat may be caused by visible freeboard height and noise generation above and below the water surface.

- Conflict with other users of the ocean, such as commercial shipping and recreational boating, is possible.

Water Turbines

Most water turbines require a constant, steady flow of water in order to generate a steady flow of electricity. Both tidal- and wave-generated energy are multidirectional and variable. This variable and cyclic operation does not produce a steady flow of electricity and can lead to low operational efficiency. All equipment used on the surface or below the surface of the ocean must be durable enough to survive storms and to survive corrosion from seawater.

Biofuels

The first generation of biofuels is being produced, on a large scale, using feedstocks, to produce ethanol. The second generation of biofuels in the development stage is being produced from non-food materials, including switch grass. Efforts are underway to develop, as a third generation, production methods for converting micro-algae to a biofuel.

First-Generation Biofuels

First-generation biofuels are made from sugar, starch, vegetable oil, or animal fats using current technology to convert these feedstocks into a combustible liquid. These feedstocks are part of the animal and human food chain. Ethanol is produced from the fermentation of sugar. Fermentation of sugar is accomplished by the addition of yeast. Glucose is the preferred form of sugar for fermentation because carbohydrates are easier to convert to glucose than cellulose. The varieties of yeast needed to ferment glucose on a commercial scale are readily available. The majority of ethanol being produced is made from corn because it contains large quantities of carbohydrates.

The production of ethanol from corn is a mature technology that is not likely to see significant reductions in production costs. It takes about 1.5 gallons of ethanol to deliver the same mileage as 1.0 gallon of gasoline. Currently, gasoline blended with ethanol cannot be shipped in pipelines. The moisture in pipelines and storage tanks is absorbed by the ethanol. Water mixed with ethanol causes the mixture to separate from the gasoline. Biofuels can cause stress corrosion cracking in standard pipeline materials.

The cost of producing and transporting ethanol will continue to limit its use as a renewable fuel. Ethanol relies heavily on federal and state subsidies to remain economically viable as a gasoline-blending component. The cost to produce and deliver one gallon of biofuel is more than a gallon of gasoline. However, the federal subsidy of fifty-four cents per gallon makes it possible for ethanol to compete as a gasoline additive. Corn prices are the dominant cost factor in ethanol production, and ethanol supply is extremely sensitive to corn prices. As an example, ethanol production dropped sharply in mid 1996 when late planting due to wet conditions resulted in short corn supplies and higher prices.

It is estimated by various sources that about 1 billion people across the world are hungry. Sixteen thousand children in our world die from hunger-related causes every day! The United States Department of Agriculture reported for the year 2007 that 36.2 million people in our country live in households considered to be food-insecure. Of these 36.2 million, 23.8 million are adults and 12.4 million are children. Is it morally right to convert food to a fuel to burn in our automobile engines when 12.4 million children in our own country are malnourished and 16, 000 children around the world are dying each day? Have we become a decadent society that burns food in order to keep our cars running, while we have children that are starving right here in our country? It would be more humane to use our tax dollars to reduce the number of starving families than spend it on ethanol subsidies.

Using natural food products for the production of biofuels is currently part of the cause for our increase in food prices. The production of biofuels usually requires the use of fossil fuels in the conversion process. The economics of using fossil fuels for the production of a biofuel, coupled with a decrease in the food supply, is definitely not an attractive alternative for reducing our use of fossil fuels.

The Biomass Research and Development Board presented a top-line biofuels-commercialization timeline in 2008.[15] Between now and the year 2022, continued research and development is needed to enhance and refine technologies by reducing costs and increasing efficiencies. A large investment of money is needed to support a proposed level of biofuel production by 2022. Hundreds of biofuel plants are needed. There will also be a need for skilled labor, construction materials, and the development of distribution alternatives. A distribution infrastructure that includes new pipelines, rail carriers, and trucks is needed. The board concluded that current production costs remain too high for biomass-based fuels to compete in the marketplace.

Second-Generation Biofuels

Second-generation biofuels are obtained from non-food crops. The non-food crops considered include stalks of wheat, corn, and wood. The non-food crops supply cellulose, a material that can be chemically converted to ethanol. The non-food crops are inedible and can be considered waste material. These plant materials do not divert food away from the animal or human food chain.

A study partially funded by the National Science Foundation by Duke University concludes that switchgrass, used for the production of ethanol, is better than corn for curbing greenhouse emissions.[16] The switchgrass, after first planting, will continue to grow each year.

It can be harvested each year for many years while trapping carbon in the soil. Corn must be planted each year with the expenditure of carbon dioxide as a result of the energy required to till the soil and plant the corn. The Duke researchers analyzed over 140 soil samples and found that conventional corn farming can remove 30 to 50 percent of the carbon stored in the soil. In contrast, switchblade-ethanol production entails mowing plants as they grow. Some of this land is already in a conservation reserve. Their analysis found for the case of switchgrass, the soil could actually accumulate carbon by between 30 to 50 percent instead of reducing the soil's carbon level, as was the case found for soil samples used to grow corn. Setting aside acreage and letting it return to native vegetation was rated the best way to reduce greenhouse gas emissions when compared to the corn-ethanol production over the next forty-eight-year period of time.

A conclusion of the Duke study suggests that until switchblade ethanol production is feasible or corn-ethanol technology improves, corn ethanol subsidies are a poor investment economically and environmentally.

Algal—The Third-Generation Biofuel

A third generation of biofuel, called *algal*, can be produced from microalgae. Microalgae are single-cell, photosynthetic organisms known for their rapid growth and high energy content. Some algal strains are capable of doubling their mass several times each day. In some cases, more than half of the mass consists of lipids or triacylglycerides, the same material found in vegetable oils. These bio-oils can be used to produce advanced biofuels including biodiesel, gasoline, and fuel for jet engines.

Biodiesel fuel consists of chemicals known as fatty acid methyl esters that can be used as a diesel fuel substitute or a diesel fuel additive. This biofuel is typically made from oils produced from agricultural crops such as soybeans or canola. It can also be produced from a variety of other feedstocks, such as animal fats. Most of the biodiesel fuel is made from soybeans. As the price of soybeans increased, biodiesel producers have been using more and more waste, animal fat, and grease from restaurants. Algae used to produce biofuels offers certain advantages over other feedstocks. For example, algae contain fat pockets that help them float in water. The fat can easily be collected from the water surface and processed into biodiesel fuel.

Algae can be cultivated in large open ponds or in closed photobioreactors located on non-arable land, such as a desert, and located in a variety of climates. During photosynthesis, algae uses solar energy to help convert carbon dioxide into biomass. The water used to cultivate the algae must be enriched with carbon dioxide. This process offers the opportunity to make productive use of carbon dioxide emissions from fossil-fueled electric power plants, biofuel production plants, and other sources. Many species of algae thrive in seawater, water from saline aquifers, or even wastewater from water treatment plants. The remaining biomass residues resulting from the production of biofuel can be burned to generate heat, used in a digester to produce methane gas, used for fermentation feedstock to produce ethanol, and as an additive in animal feed.

Algal biofuels are not economical to produce within the state of our technology. If algae fuels were produced in large volumes, it is estimated that the fuel would cost more than $8.00 per gallon. For comparison, biofuels produced from soybean oil costs about $4.00 per gallon. It is anticipated that production costs can be reduced if research and development continues over the next five to ten years.

Geothermal Energy

There are two sources of stored energy in the earth being used to produce electric power or provide heat for residential and commercial buildings. The first source involves extracting steam or very hot water from beneath the earth's surface. The stored thermal energy, in steam or hot water, is called *hydrothermal energy*. The heat energy can be converted to produce electricity or to supply heat for buildings located in cool climates. The number of underground locations that yield steam or hot water is very limited.

Deep underground, hot dry rocks are found to be enormously more prevalent when compared to reservoirs of underground hot water and steam. The second geothermal energy source, called *enhanced geothermal energy*, was developed to extract thermal energy from hot, dry rocks found below the earth's surface.

Geothermal energy has provided commercial base-load electricity around the world for more than a century. However, it is often ignored in national projections of our future energy supply. A widespread misperception is that the total geothermal resource is associated with the need for high-grade hydrothermal energy.[17] The high-grade systems are too few and too limited in their distribution in our country and will not appreciably supply our future energy needs. The misperception has led to undervaluing the long-term potential of geothermal energy. We have missed an earlier opportunity to develop the technologies for obtaining reusable energy from large volumes of accessible hot, dry rock formation most prevalent in the western half of our country. Enhanced geothermal energy plants have several positive attributes when compared to most other renewable energy sources. Plants can be located over a wide range of geographic locations in our country. Unlike the wind farm, they provide a base-load capability and no energy storage infrastructure is needed. Enhanced geothermal plants produce negligible emissions and each plant requires a small area of property when compared to an array of solar-electric modules.

Hydrothermal Resources

Conventional geothermal technology involves the production of useful energy from natural sources of steam, or much more commonly, hot water. Hydrothermal resources like Old Faithful located in Yellowstone National Park, Wyoming, are found in a number of locations around the world but are few in number. In most places, the earth grows hotter with increasing depth, but mobile water is absent.

Hydrothermal resources are used effectively for both electric and non-electric applications in our country. Most operational experience has been with very small electric geothermal plants operating in a range between 10 and 300 megawatts.[18]

A company based in Provo, Utah, has started developing two new geothermal projects, bringing its total number of projects under development to seven. The new projects include a 10-megawatt plant in Utah and the company's first ten-megawatt plant in Oregon. California has installed about 86 percent, or 2,500 megawatts, of overall geothermal power generation capacity in our country. Nevada represents about 11 percent, or 318 megawatts, of installed geothermal electric power. California has the greatest potential for geothermal energy with about 11,000 megawatts of undiscovered geothermal resources, as estimated by the United States Geological Survey.

There are seven states generating electricity with geothermal energy, including Idaho and New Mexico. Oregon and Wyoming are working on geothermal plants planned to be in operation between 2011 and 2014. Utah currently has two geothermal electric generation power plants. One is located in the southwest part of the state, while the other is located in central part of Utah. Nevada has in development between 1,100 to 1,900 megawatts of geothermal capacity. California has in development between 900 to 1,000 megawatts of geothermal capacity. Geothermal projects in development during 2009 represent a maximum capacity of about 3,900 megawatts.

Enhanced Geothermal Systems Using Hot Dry Rock Resources

Like hydrothermal energy resources, hot dry rock holds the promise for being an environmentally clean energy resource. Carbon dioxide emissions can be expected to be practically zero.

The Fenton Hill Story

During 1970, a team of scientists and engineers at Los Alamos National Laboratory developed a plan to access the hot dry rock resource and bring its contained heat to the surface for practical use. The development concept entailed drilling a well into hot crystalline rock. Water under pressure could create a large, vertical, disc-shaped fracture in the hot rock. A second well would be drilled to access that fracture at some distance above the first wellbore. The system would be operated by:

- injecting pressurized cold water through the first well into the deeper part of the fracture
- flowing water across the hot surface of the fracture.
- returning the water to the surface as superheated fluid through the second wellbore

After extracting its useful energy, the same water would be re-circulated to mine more heat. Larger systems could be developed by creating multiple fractures spaced along a single set of wellbores inclined toward the horizontal at depth. This original concept was significantly modified as researchers learned more about the characteristics of the engineered geothermal reservoirs created during hydraulic fracturing operations.

After a number of preliminary drilling and fracturing experiments, a site at Fenton Hill, New Mexico, was chosen for the establishment of the world's first hot dry rock circulation system. Fenton Hill is located about forty miles west of Los Alamos, and the land is owned by the US Forest Service.

Phase I System: 1974 to 1980

The Phase I project at Fenton Hill started in 1974 and was completed with a series of tests by 1980. The first deep borehole was drilled into a hot rock formation to a final depth of about 9,600 ft. The bottom hole temperature was measured at 365 degrees Fahrenheit. The borehole was pressurized with water to cause a fracture in the rock formation at the bottom of the hole. It was first thought that the fracture would open below the bottom of the first borehole. A second borehole was directionally drilled beneath the bottom of the first borehole with the intent to intersect with the hydraulically induced fracture.

A flow test was conducted by pumping water into the first drilled hole and measuring the water flow out of the top of the second borehole. It was found that only a seepage flow of water was being expelled from the second hole at ground level. The first borehole was pressurized with water again to increase the volume of the fractured rock. Another flow test was conducted that resulted in very little change in the volume flow when compared to the first test. A re-drilling of the first borehole was conducted in mid 1977 that finally succeeded in producing a satisfactory flow from the first hole to the second. The connection was the first-ever fracture connection between two boreholes in deep hot rocks and is considered the world's first hot dry rock reservoir.

During the final flow test of the Phase I reservoir in 1980, the temperature of the produced fluid declined from an initial value of 313 degrees to 300 degrees Fahrenheit at a near constant flow rate of ninety gallons per minute and an injection pressure of 1,200 pounds per square inch. The scientific data and engineering experience acquired during testing of the Phase I research reservoir provided the basis for the development of the larger, hotter, and deeper Phase II engineering-scale hot dry rock system.

Phase II System: 1979 to 1992

In 1979, construction of the Phase II hot dry rock system was initiated. The results of the Phase I project provided little reason to doubt the validity of the original Los Alamos concept. Phase II began by attempting to generate a series of vertically fractured reservoirs. The deepest wellbore penetrated to a depth of about 14,000 feet, with the last 3,300 feet inclined from the vertical at an angle of 35 degrees. The second wellbore was drilled to a total depth of about 13,000 feet, with the last 3,300 feet angled at 35 degrees from the vertical and positioned above the sloped portion of the deeper wellbore. Between 1982 and 1984, numerous hydraulic fracturing operations were conducted at several points along the sloped portion of the wellbore. Initial fracturing operations failed to connect the two wellbores. Advances in seismic science at the time made it possible to more accurately locate the origins of the micro-earthquakes generated during the hydraulic fracturing. With this information, researchers were better able to determine where the reservoir fractures were located and how they were extending.

The most extensive hydraulic fracturing operation was conducted in the lower wellbore at a depth of about 11,700 feet. About 5.7 million gallons of water at a surface pressure of about 7,000 pounds per square inch was injected into the lower wellbore. Seismic data indicated that the reservoir created during this operation was developing in a three-dimensional manner. The reservoir took the shape of a 300-foot-thick ellipsoid, with the longest axis oriented along the trajectory of the wellbore. Even though the size of the reservoir was increased, a connection between the two wellbores was not established. It was decided to re-drill the lower portion of the upper wellbore to penetrate the newly formed reservoir. By re-drilling and a small amount of re-stimulation in the re-drilled wellbore, a number of hydraulic connections between the two wellbores were established. With this accomplishment, the Phase II reservoir was ready for testing.

The most meaningful definition of reservoir volume is the flow-accessible or heat-transfer volume. The heat-transfer volume includes only those portions of the reservoir that are accessible to the circulating fluid. The term *fluid* is used because a fluid can be either a liquid, in this case water, or a gas, such as superheated steam. From a practical point of view, the heat transfer volume is the only part of the reservoir that can provide energy to the circulating water and, ultimately, to the energy-production facility at ground level. The heat-transfer volume for the reservoir was estimated at about 1.6 to 2.1 billion gallons.

The larger fluid-accessible volume of the reservoir encompasses all parts of the reservoir, even dead-end joints that are reached by the injected fluid. The fluid-accessible reservoir volume was estimated to be about three times larger the flow-accessible volume. Portions of the larger volume can be used to increase the productive capacity by adding more wells.

The original Los Alamos team concept that hydraulic pressure would cause competent rock to rupture and create a disc-shaped fracture was refuted by the seismic evidence found during Phase II. Instead it was newly understood that hydraulic stimulation leads to the opening of existing natural joints in the underground hot rocks that have been sealed by secondary mineralization.

With the Phase II reservoir and wellbores in place, work on the design and construction of a surface plant began in 1987 and was completed by 1991. The newly constructed surface plant allowed the reservoir to be flow-tested in a manner simulating the operation of a commercial hot dry rock facility. The heart of the plant was the injection pump. The unit provided the pumping power to force the water down the injection wellbore, across the reservoir, up the production wellbore, and back again to the inlet. Surface piping delivered the circulating water to a heat exchanger, which cooled the water to ambient temperatures. Engineering information was recorded during operation, which included water temperatures, pressures, and flow-rate. The data was automatically measured and recorded at frequent intervals, and safety systems were in place to assure that the plant would shut down in the event of certain pre-selected operating values being outside of a safe range. It was found entirely feasible to operate the plant for extended periods of time with no need for on-site personnel. The ability to operate a plant with little or no need for on-site personnel has important economic implications for commercializing hot dry rock technology.

Phase II long-term flow test results include these findings:

- The rapid attainment of repeatable operating conditions indicated that the reservoir had the stability required for predictable energy production.

- The reservoir has the capability to change the production rate to produce more energy during periods of peak demand when power is most valuable.

- The reservoir could have operated for many years without an appreciable decline in the temperature of the produced fluid.

- Water loss in the fluid loop declined with time at constant reservoir pressure.

- Concentrations of dissolved solids reached about 3,500 parts per million in the closed fluid loop. That level is about one-tenth the salinity of seawater.

- Dissolved gases reached an equilibrium level of 2,000 parts per million in the closed fluid loop, with carbon dioxide being the preponderant gas.

- Four to six megawatts of thermal power was produced.

- At the temperature produced, a little less than 0.5 megawatts of electricity could have been generated had the facility been equipped to convert the thermal energy to electricity.

- Treated sewage can be used as a source for feed water. Under high temperatures and pressures within the reservoir, the water would be sterilized. Purified water, as well as thermal energy, can be recovered at the surface.

- Seawater can be desalinated, provided proper measures were in place to handle the large volume of salts that would be returned to the surface along with the superheated steam.

Three major issues must be resolved before the hot dry rock systems become a significant contributor to the commercial energy market. The issues are:

- Reservoirs must be able to produce an economic rate of return in relation to the investment.

- The lifetime of a reservoir must be sufficient to warrant the large initial investment required to establish a hot dry rock system.

- The reservoir at Fenton Hill must be established as a routine site rather than an anomaly.

Research and development efforts, both in our country and others, have made a significant start toward resolving the three issues. Implementation is now essential. What is most needed is a hot dry rock facility that produces energy for market. This facility can build the track record that will make this technology an attractive investment to power producers in our country and around the world. Programs underway in both Europe and Australia show promise of developing the first commercially viable hot dry rock facility. The hot dry rock system can move rapidly toward becoming a major clean energy resource of the twenty-first century. A cumulative capacity of 100,000 megawatts can be achieved in our country by tapping into enhanced geothermal systems over the next fifty years, according to a study released by the Massachusetts Institute of Technology in January 2007.

Enhanced geothermal system technology has advanced since its infancy in the 1970s at Fenton Hill, New Mexico. Field studies have been conducted worldwide for more than thirty years. The studies show that the technology is feasible in terms of producing net thermal energy by circulating water through stimulated regions of rock at depths ranging from two to three miles. The enhanced geothermal system can easily be deployed to heat and provide electric power to large buildings without the need for energy storage on-site. For other renewable options such as wind and solar photovoltaic, the ability to provide heating and electricity are not as attractive. The size of the enhanced geothermal system is easily variable from 1 to 50 megawatts.

Like all new energy-supply technologies, state and local policies are needed to allow the entrance of geothermal energy to compete in the evolving electricity markets. These policies must be similar to those granted to the oil, natural gas, and other mineral-extraction operations over the past years. Provisions are needed for accelerated permitting and licensing, loan guarantees, depletion allowances, intangible drilling write-offs, and accelerated depreciations, as well as those policies associated with cleaner and renewable energies such as production tax credits, renewable credits, and portfolio standards. Such an approach would parallel the development of the coal bed methane industry in our country.

Data indicates that geothermal plants are far more environmentally benign than the other conventional plants. For example, the enhanced geothermal system would not produce carbon dioxide, sulfur dioxide, or oxides of nitrogen emissions. The emissions of particulate matter would be negligible. Water mixed with chemicals is required for well drilling to provide cooling for the drilling bit and rock chip removal. The mixture of water and chemicals is re-circulated after being cooled and strained. Makeup water is required to compensate for evaporation losses during cooling. Liquid streams from well drilling, water injection below ground level, fracturing underground hot rock formations with water under high pressure, and geothermal operation may contain a variety of dissolved minerals. The amount of dissolved minerals increases significantly with the temperature of the hot rocks, usually above 380 degrees Fahrenheit.

Some of the dissolved minerals could be toxic, such as boron and arsenic, which could poison the surface or ground waters and also harm local vegetation. Liquid streams could enter the environment through surface runoff or through potential breaks in the well casing. Surface runoff is controlled by directing fluids to impermeable holding ponds and by injection of all waste streams deep underground. To guard against fluids leaking into shallow, fresh-water aquifers, well casings are designed with a pipe within a pipe. The concentric pipes provide a redundant barrier between the inside of the well and the adjacent earthen formations that vary with depth. There is little chance for surface contamination during plant operation because all the produced fluid is re-injected.

For the case of noise pollution, the highest noise levels are produced during the well drilling, stimulation, and initial testing phases.

During these operations, noise levels may range between 80 and 115 decibels. During normal operation of a geothermal power plant, noise levels vary between 71 to 83 decibels when measured at a distance of about 3,000 feet away from the plant. For comparison, congested urban areas typically produce noise levels between 70 to 85 decibels. Near a major freeway the level is about 90 decibels. And jet airplanes, during take-off, produce noise levels in excess of 120 decibels.

The area of land needed for a geothermal plant is very small compared to a solar-thermal plant or a solar photovoltaic plant. The amount of land needed for the solar-thermal plant requires about twenty times more area than a geothermal plant per megawatt. The amount of land required for a solar-photovoltaic plant is about fifty times more than the geothermal plant per megawatt. Because the surface area needed for geothermal plants is small compared to solar electric plants, problems related to loss of habitat or disturbance of vegetation would be relatively minor or nonexistent at future hydro-thermal projects in our country.

The development of a geothermal field may require tree and brush removal. Clearing of the land is needed for installation of equipment including a powerhouse, an electric substation, piping, and an emergency holding pond. Once the plant is complete, the topography surrounding the plant can be restored to a semblance of its original natural appearance. Geothermal plants generally have a low profile and are much less conspicuous than for example wind turbines, solar towers, or coal-fired electric plants with chimneys as tall as 490 to 650 feet. Heat rejection into the atmosphere from a thermal plant is higher per unit of electricity produced than for fossil fuel and nuclear plants. Considering only thermal discharges at the plant site, a geothermal plant rejects two to three times more heat than a nuclear power plant of comparable power rating. On balance, considering all the technologies available for generating large amounts of electricity together with their associated environmental impacts, the enhanced geothermal systems should be made a predominate part of the mix of renewable energy sources.

In December 2008, the Department of the Interior's Bureau of Land Management published the Record of Decision and Approval Management Plans Amendments for geothermal leasing in the western United States. The Record of Decision allocates about 111 million acres of public lands as open for leasing under bureau management. An additional 79 million acres of National Forest System lands are also legally open for leasing. About 90 percent of our geothermal resources are found on Federal lands.

Lands within a unit of the National Park System, such as Yellowstone National Park, which features a number of geothermal geysers, will continue to be unavailable for leasing. Federal lands in the west contain the largest supply of known resources of geothermal energy in the country. Geothermal leasing revenues and royalties, in California and Nevada, are shared with these states and counties where the leases are located, with 50 percent going to the state and 25 percent to the county.

With regard to the issues on construction, the availability of geothermal drilling rigs has improved since 2008. The supply of drilling rigs remains an issue during 2010 through 2011, when the majority of projects will reach the feasibility stage. Another bottleneck facing the industry is the availability of, and delivery time for, geothermal turbines. The demand for these types of turbines is already high. With current development plans, the demand is expected to increase sharply by 2011 until 2013. During this time period, most projects now in development will reach the design and construction stage. Another major obstacle facing geothermal energy utilization in our country is the shortage of educated and trained people. Trained and experienced people are needed to further develop and carry out the technology of geothermal plant locations, design, and construction. State and local governments, industries, and universities should recognize this need and offer opportunities for people to support the research and technology needed to expand geothermal energy to meet the potential that it offers.

To augment our country's geothermal technology efforts, an International Partnership for Geothermal Technology was formed in 2008. The partnership between the United States, Australia, and Iceland commits the three countries to aggressively foster and promote cutting-edge geothermal technologies to promote energy security and address global climate change. The three countries each possess individual strengths and experience in geothermal energy. Together this partnership will allow each country to capitalize on a collaborative international endeavor.

A positive force working in favor of geothermal energy and other renewable sources is that many states and the federal government are adopting renewable portfolios standards, in which electricity suppliers are required to buy 25 percent of their electricity supply from renewable energy sources. Enhanced geothermal systems have the potential to be the world's only ever-present form of base-load renewable energy. Recall that solar power, wind power, and tidal- and wave-action renewable energy sources provide periodic bursts of electricity. Geothermal power plants can be relied on to supply a steady flow of electricity and can be counted on to supply base-load power.

Energy Conservation

The Department of Energy manages a Building Technologies Program and an Industrial Technologies Program.[19]

The Building Technologies Program has five subprogram activities:

- Residential Building Integration
- Commercial Building Integration
- Emerging Technologies
- Technology Validation and Market Introduction
- Equipment Standards and Analysis

The Industrial Technologies Program has two subprogram activities:

- Industries of the Future (Specific)
- Industries of the Future (Cross-cutting)

The Building Technologies Program

The mission of the Building Technologies Program is to change the landscape of energy demand by decreasing energy usage in homes and buildings. The program brings together science, discovery, and innovation to develop the technologies, techniques, and tools for making residential and commercial buildings more energy efficient. Making homes and buildings more energy efficient means the energy consumption of homes and buildings will be reduced.

Buildings account for more than 70 percent of the electric energy consumed in our country. The building sector in our country is responsible for 38 percent of the total carbon dioxide emissions. Efficiency gains will reduce the energy demands of commercial buildings and residential homes and therefore reduce the consumption of electricity generated from the combustion of fossil fuels. Improved energy efficiencies will in turn reduce carbon dioxide emissions. Space-heating accounts for the primary use of energy in homes. In certain parts of the country, homes are heated exclusively by natural gas or petroleum derivatives such as kerosene. For new homes designed for, or existing homes modified for, higher energy efficiency, a reduction in the need for petroleum derivatives and natural gas can be realized.

Residential Buildings Integration

The Residential Buildings activity focuses on improving the efficiency of the approximately 1.5 to 2.0 million new homes built each year and over 100 million existing homes. Efficiency improvements can be accomplished via research, development, and technology-transfer activities. The program seeks to make improvements by using a systems engineering approach to optimize the energy-efficiency technologies in the whole building. The program also seeks to ensure the health and safety of the buildings and to integrate renewable technologies into the buildings.

Commercial Buildings Integration

The Commercial Buildings Integration activity addresses energy savings opportunities in new and existing commercial buildings. The research, development, and demonstration of whole building technologies include sensors and controls, design methods, and operational practices. These efforts support the DOE zero-energy building goal to reduce building energy needs and to develop design methods and operating strategies that incorporate solar and other renewable technologies into commercial buildings.

Emerging Technologies

The Emerging Technologies activity conducts research, development, and technology transfer associated with energy-efficient products and technologies for both residential and commercial buildings. Efforts address high-impact opportunities to reduce energy usage of equipment located inside of buildings. The equipment types include lighting and building materials within the buildings such as windows, solar heating, and cooling.

Technology Validation and Market Introduction

The Technology Validation and Market Introduction activity will accelerate the adoption of clean, efficient, and domestic energy technologies. Three major initiatives within this subprogram are the Energy Star activity, the Rebuild America activity, and the Building Energy Codes.

- Energy Star is a joint Department of Energy and Environmental Protection Agency activity designed to identify and promote energy efficient products.

- Rebuild America is aligned with the Commercial Building Integration activity to accelerate the adoption of advances in integrated commercial building design, software tools, practices, advanced controls, equipment, and lighting. The activity will target companies in the retail, lodging, restaurant sectors, commercial property developers, owners, and operators, as well as in the school and hospital sectors.

- Building Energy Codes submits code proposals and supports the upgrades of the model building energy codes. The activity provides technical and financial assistance to states to update, implement, and enforce their energy codes to meet or exceed the model codes. Section 304 of the Energy Conservation and Production Act, enacted in 1976, calls for this activity. It also supports the standards for manufactured housing required by Section 413 of the Energy Independence and Security Act of 2007.

The DOE states that reduced energy use in buildings can be expected to lead to reduced energy bills for American families and businesses. New technologies developed with the help of the DOE building technologies program and with the participation of domestic industries will create jobs, spur economic growth, and restore America's role as a global innovator and exporter of high-tech products. Efficient buildings have the added benefit of mitigating the need for the electric power industry to construct expensive new power plants. The need for fewer new plants will save power companies money. The savings will flow directly to the electricity consumer. Power company savings might be used to modernize the electric grid and modernize other energy infrastructure investments.

The Department of Energy is working to improve the energy efficiency of our nation's buildings through new technologies and better building practices.[20] Energy-efficient buildings improve the lives of Americans by saving consumers money, lessening our demand for fossil fuels, decreasing the need for new power generation, and reducing environmental emissions. The Department of Energy provides the Building Codes Program, an informational resource on national model energy codes.

Strengthening energy codes increases the likelihood of energy and cost savings in new construction and renovations to existing buildings. New buildings can be designed to be both more comfortable and more energy efficient, and it is possible to reduce heating and cooling costs by about 50 percent when compared to older existing buildings. For existing buildings, the most cost-effective time to improve the energy efficiency of these buildings is when the building must be renovated for new owners or tenants. At that time, energy efficient insulation, heating and cooling equipment, and lighting can be installed. The DOE works closely with the International Code Council, the American Society of Heating, Refrigeration, and Air-conditioning Engineers, the Illuminating Engineering Society of North America, and other groups that use standard codes for their design and construction of buildings. Together, these groups and the Department of Energy are developing more stringent and easier-to-understand building energy codes. These groups are also identifying candidate energy-efficient technologies and practices and working to remove barriers to these technologies in the national model energy codes.

The DOE provides direct financial and technical assistance to states to promote the adoption, implementation, and enforcement of state and local building codes. They help states modify national model energy codes to meet state needs and develop state-specific code compliance software and training materials. The DOE also provides code compliance training within states and analyzes the energy and economic impact of state and local building codes. The results of these activities are made available to our states and local governments. The software helps our states to modify national energy codes to meet state needs and develop state-specific compliance software and training materials. It also provides code compliance training within states and analyzes the energy and economic impact of state and local building codes.

The DOE efforts to improve the energy efficiency of new buildings and homes have improved the energy efficiency of nearly 3 billion square feet of new commercial floor space and nearly 4 million new households. The Building Energy Codes Program is helping to continue this improvement in new home and building energy efficiency.

The DOE develops and distributes compliance tools and materials for designers, builders, product manufacturers, and code officials to comply with standard energy codes. The compliance tools and materials include software and accompanying user guides, videos, CDs, training materials, and compliance manuals. The DOE sponsors informational workshop and training events. A building energy code website is a major source of communication and information exchange on codes for buildings. Users can download the compliance software and training materials free of charge. A technical staff, supported by the DOE, answers questions on energy codes, provides information and technical assistance on energy codes for buildings, and connects inquiries to other resources.

A block grant program is being administered by the Office of Weatherization and Intergovernmental Programs in the Office of Energy Efficiency and Renewable Energy of the US Department of Energy.[21] The purpose of the block grant program is to provide funding to units of local and state governments, Indian tribes, and territories to develop and implement projects to improve energy efficiency, to reduce energy use, and to reduce fossil fuel emissions in their communities. Over $2.7 billion in grants are available for theses communities under the American Recovery and Reinvestment Act of 2009. By mid 2009, the Department of Energy had awarded more than 1,000 Energy Efficiency and Conservation Block Grants, totaling over $1 billion.

In September, 2009, in order to initiate the block grant program, the DOE submitted a request for information seeking feedback from stakeholders on a planned competitive Funding Opportunity Announcement in support of the American Recovery and Reinvestment Act.[22] The DOE anticipates that a total of up to $453 million will be available for competitive grants.

The goal of the planned competitive Funding Opportunity Announcement is to stimulate activities that move beyond traditional public awareness campaigns, program maintenance, demonstration projects, and other "on-off" strategies and projects. Rather, the DOE seeks to stimulate activities and investments by a variety of other means.

- Verify energy savings from a variety of localized projects with particular emphasis on efficiency improvements in residential, commercial, and public buildings.

- The Building Energy Codes Program provides direct technical assistance upon the request of states and local jurisdictions to help them to adopt, implement, and enforce building codes.

- Applicant localities are required to provide a viable plan to sustain the energy-efficiency efforts beyond the grant monies and grant period. Applicant localities should serve as examples of comprehensive community-scale energy-efficiency strategies that could be replicated in other communities across the country.

The DOE is specifically targeting funds for a small number of high–profile, high-impact awards that will enable large-scale programs of ongoing energy-efficiency retrofits on residential, commercial, and public buildings in geographically focused areas. The programs are intended to result in high-quality retrofits that lead to significant efficiency improvements to a large fraction of buildings within targeted neighborhoods or communities. The retrofits—which include building modifications and lighting, heating, and cooling installations—must reduce the monthly operating costs of the buildings, including any repayments of loans.

A key goal of the block grant program is to demonstrate that a level in the number of building modifications can be reached within a community to convince non-participants to invest on their own in energy-efficient modifications to their buildings and homes.

Equipment Standards and Analysis

The goal of the Equipment Standards and Analysis subprogram is to develop minimum energy-efficiency standards that are technologically feasible and economically justified. The activity will lead to improved efficiency of appliances and equipment by conducting analyses and developing standards. Technical feasibility and economic justification is contained in the Energy Policy and Conservation Act, enacted in 1975, as amended. Appliance and equipment standards help to reduce our energy consumption.

The DOE continues to take steps consistent with federal laws to implement legally required energy-efficiency standards to meet the judicial and statutory deadlines. Authorization for these activities is found in the Energy Policy Act of 2005 and the Energy Independence and Security Act of 2007. The Equipment Standards and Analysis subprogram will continue rule-making for various product categories, some of which are listed below.

Table 1: Partial List of Commercial Products Being Evaluated by the DOE to Meet Energy Efficient Equipment Standards

Residential Water Heaters	Direct Heating Equipment
Small Electric Motors	1 to 500 hp Electric Motors
Pool Heaters	Clothes Dryers
Fluorescent Lamp Ballasts	Room Air Conditioners
Central Air Conditioners	Heat Pumps
Furnas Fans	Lighting Products
Battery Chargers	Residential Refrigerators
Residential Clothes Washers	Commercial Refrigeration Equipment

The DOE activities during FY 2010 included responding to waiver requests from manufacturers and requests for input and recommendations to the DOE Office of Hearings and Appeals. The Department of Energy will initiate energy conservation standard rule-makings on furnace fans, 1 to 500 hp electric motors, and commercial refrigeration equipment. In February, 2009, President Obama issued a Memorandum to the Secretary of Energy requesting that the DOE "work to complete prior to the applicable deadline those standards that will result in the greatest energy savings." As a result, the DOE is considering energy rule-making for additional products that may include televisions, commercial automatic ice-makers, and certain plumbing products.

The external factors as listed below could affect Building Technologies' ability to achieve its strategic goal.

• Fragmented Construction Market.

There are several factors that can hinder the private sector to make investments in energy-efficient building technologies. Theses include a highly diversified industry comprised of thousands of builders and manufacturers, none of which has the capability to sustain research and development activities over multi-year periods.

• Communication Between Professional Groups.

The compartmentalization of the building professions do not typically apply integrated strategies for selecting sites or for the construction, operations, and maintenance. The building professions include architects, designers, developers, construction companies, engineering firms, and energy services providers.

• Upfront Costs

The high initial cost of energy-efficient building appliances can keep consumers from purchasing them even if they are cost-effective in the long run.

- The Housing Market

Conditions in the housing market that would affect the number of new subdivisions being built would slow down research on zero-energy buildings. The last phase of research is having a builder construct a subdivision using technologies developed by the building technologies program in order to prove them in a real-world setting. If fewer subdivisions are being constructed by more risk-adverse contractors, it could slow the building technologies research considerably.

- Unit Price of Renewable Energy.

Zero-energy building goals are contingent upon the development of cost-effective, small-scale renewable energy systems.

Industrial Technologies

The mission of the Industrial Technologies Program is to significantly reduce the intensity of energy use by the industrial sector. The DOE will continue its research development and demonstration of the next generation of manufacturing technologies. The Industrial Technologies Program is leading the Federal Government's efforts in industrial energy efficiency, leveraging the knowledge and expertise of the National Laboratories and broadening private sector partnerships. The program's activities help our nation's industries advance their global competitiveness, keeping jobs in America and reducing our reliance on imported oil and other goods while abating greenhouse gas emissions.

Some of the challenges to the program include external factors that could affect the program's ability to achieve its goals. The factors are listed below.

- Industries must maintain their economic health and profit margins.

- The rate of market growth and technology adoption can vary.

- Labor and material costs, capital investment requirements, and the cost for improvements through new technology must be considered.

- Foreign competition will be a deciding factor for technology improvements.

- Energy supply availability and the cost of energy can change the decision for technology upgrades.

- Safety and environmental regulations and environmental policies at the national and state level could affect the acceptability of new technology implementation. For example, federal efforts to reduce carbon and other emissions may affect the choice of energy sources.

Industries of the Future (Specific)

The Industries of the Future (Specific) subprogram supports cost-shared research, development, and demonstration of advanced technologies to improve the energy and environmental performance of America's most energy-intensive industries. The subprogram partners with industries in our country to develop solutions to their top technological challenges. The industries in partnership with the DOE include the Forest and Paper Products Industry, Steel and Metal Casting Manufacturing, and the Aluminum and Chemical Industries. These industries are critical to the nation's economic prosperity and national security. The subprogram has an excellent track record of moving innovative energy-efficient technologies to the eventual introduction to their respective industrial markets.

Some of the specific activities addressed during FY 2010 were directed to transform, through research, development and demonstration of the next-generation manufacturing technologies that eliminate energy-intensive steps. The manufacturing activities include:

- Coke-less iron-making, blast furnace optimization, transformational iron-making processes, and thermal energy recovery in high-temperature steel furnaces.

- Developing a carbon-neutral pulp mill by decreasing or eliminating fossil fuels in pulping operations.

- High-efficiency water removal for pulp

- Micro-channel reactors for producing high-value chemicals and hybrid membrane/distillation technologies for chemical production.

- Development of low- to zero-carbon processes such as carbon-neutral pulping to assist domestic industries.

- Research, development, and demonstration on new high-yield and low-waste methods of manufacturing commodities including chemicals, metals, metal parts, and other components for industries such as automobile manufacturing.

The DOE continues to support the American Forest and Paper Association and the American Iron and Steel Institute. Collaborative activities include the continuation of cost-shared research, development, and demonstration as well as the utilization of new-and-improved energy technologies, industrial energy-efficiency tools, and energy management best practices.

Industries of the Future (Cross-Cutting)

Activities under the heading Industries of the Future (cross-cutting) provides the means for developing energy-saving technologies with broad benefits across a wide span of industries. New energy-saving technologies can be developed and demonstrated that will span multiple industries not within practical development reach of an individual industry. Energy-saving technologies used across multiple industries will provide economic, environmental, and energy-saving benefits. The DOE continues to work with the National Laboratories, academia, industrial companies, and equipment suppliers across many industries in an effort to demonstrate visible benefits.

During FY 2010, the DOE planned to carry out several projects as described below:

- Accelerate the adoption of combined heat and power in industry to improve energy efficiency. Specific activities will include the development of dual alternative fuels for turbines and engines that meet the stringent gas emissions regulations set in Southern California. Development of thermally activated technologies such as heat pumps and absorption cooling, or refrigeration, will assist to reduce energy usage by food processing plants and computer data centers.

- Assist companies across the country to address and identify affordable energy saving and carbon reducing opportunities in their plants.

- Assist in developing innovative new materials that can be used to make more durable manufacturing equipment and new high-value products. Continue to develop nano-composites and nano-coatings, materials for energy systems, advanced material solutions for waste energy recovery, and refractory materials for industrial systems. New activities will include energy-efficient methods for manufacture of carbon fiber composites at reduced energy and cost.

- Focus nano-manufacturing and other inter-agency manufacturing research, development, and demonstration activities. The focus will assist to enable industrial processes to build on scientific discoveries from National Laboratories and the DOE, including the mass production and application of nano-scale materials, structures, devices, and systems.

- Conduct fuel and feedstock flexibility activities leading to the development and adoption of alternate fuel and feedstock (biofuels) technologies to reduce reliance on imported oil. Seek to displace industrial petroleum and natural gas use through a targeted application-focused technology development and demonstration initiative. The initiative will help to link industrial users with advanced fuel development activities at the DOE. The activity will assist industry in integrating biofuels into manufacturing processes, improve fuel flexibility to reduce the damaging effects of fossil fuel price-hikes, and facilitate the manufacture, handling, and processing of alternative feed stocks.

Energy-Intensive Process Research and Development

The activity capitalizes on the institutional knowledge and expertise of the National Laboratories, builds cross-lab teams with appropriate industry partners, and leverages industry resources to exploit opportunities. The activity supports multi-industry research and development in four platform areas, as listed below:

- Waste energy minimization and recovery that includes high-energy steam generation and improved energy-recovery technologies.
- Industrial reaction and separation including advanced water removal.
- High-temperature processing including low-energy, high-temperature material processing.
- Sustainable manufacturing including near net shape casting and forming.

The activity will address energy savings opportunities that will benefit a broad set of industries that include food and beverage, computer, electronic, and fabricated metal products. Industrial Technologies research and development is planned to continue beyond 2015.

Part Two

Fossil-Fueled and Nuclear Electric Power Plants

Clean Coal Electric Power Plants

At the 1998 level of energy use, it has been estimated that there is still at least 50 years of oil available, 200 years of natural gas, and a whopping 2,000 years worth of coal. Our geologists predict that worldwide oil production is likely to peak sometime between now and 2040, thereby driving oil prices through the roof. The critical issue for climate, however, is not when production of fossil fuel peaks, but its global capacity. The upshot is that we may run into serious climate problems long before we run out of fossil fuels.

Carbon Dioxide Capture and Storage

The Nobel-Prize-winning Intergovernmental Panel on Climate Change recently concluded that global emissions of carbon dioxide must be cut by 50 to 80 percent by 2050. Such a reduction of carbon dioxide emissions is deemed necessary to avoid the most damaging effects of climate change. We are told that fossil fuels will continue to be needed between now and the year 2030. Renewable energy production during the next twenty years is not expected to be sufficient to replace the energy provided by fossil fuels.

Part of our nation's energy plan is to develop the technology to extract carbon dioxide emissions from coal-fired electric plants.[23] The extracted carbon dioxide gas would then be compressed, thereby turning the gas into liquid carbon dioxide. The liquid carbon dioxide would be injected under pressure into underground rock formations.

The American Recovery and Reinvestment Act legislation, authorized by Congress in 2009, devoted about $2.4 billion for pilot projects that would further this emerging technology for possible future applications. The pilot projects are intended to determine the technical, economic, and environmental feasibility of this approach in an effort to reduce the carbon dioxide emissions from coal-fired plants.

Several industries, including coal, electric energy, oil, automobile, and our railroads, support the concept of extracting carbon dioxide from emissions produced by the combustion of coal. The technology being developed, called "carbon dioxide capture and storage," is aimed at capturing and storing the carbon dioxide emissions from industrial plants and coal-fired electric plants.

Capture and storage of carbon dioxide is planned to bridge the time period, while we are dependent on fossil fuels, from now to the year 2030. It is anticipated that by 2030, we can reduce our need for fossil fuels and be able to produce energy with very low emissions of carbon dioxide. The technology faces many challenges including technical, regulatory, and economic issues. The largest challenge is how best to capture and safely store huge amounts of carbon dioxide. Large-scale pilot and demonstration projects are planned to show that the process will work. The demonstration plants require substantial government and private investment. The plants will help to identify ways to decrease the cost of energy produced from clean coal combustion. The demonstrations will help to identify the appropriate technologies, equipment, and the skills needed for the production of clean energy from coal-fired combustion. In addition to the technical and cost challenges, a regulatory framework is needed to clarify long-term rights, liabilities, and technical requirements. Further policy changes are essential, like the grants and subsidies currently allowed for the installation of wind and solar energy devices to support carbon capture and storage.

Coal-Fired Electric Power Plants

Coal's share of total net generation of electricity continues to decline from 52.8 percent in 1997 compared to 48.5 percent in 2007. Net generation from natural-gas-fired electric power plants increased during 2007 alone by 9.8 percent. The gas-fired natural gas plants contributed 21.6 percent of the total net generation of electricity in 2007 and surpassed the generation contributed by nuclear electric power plants. For many years, nuclear generation historically was the second leading source of total net generation after coal until 2007.

In 2002, our country emitted a total of 6,183 million tons of carbon dioxide from the combustion of all fossil fuels including coal, oil, petroleum, and natural gas. We have about 500 coal-fired electric plants in operation. Our coal-fired electric plants emitted 2,058 million tons of carbon dioxide in 2002. Gasoline-powered automobiles in our country produced 1,296 million tons of carbon dioxide during the same year. Together our coal-fired electric plants and gasoline-powered automobiles produced over 50 percent of the total carbon dioxide emissions during 2002. If our goal is to reduce carbon dioxide emissions by 50 percent, it is clear that a reduction of coal-fired and automobile emissions have to be substantially reduced.

During the combustion of coal, the carbon contained in coal combines with oxygen to form carbon dioxide. For every pound of coal burned a little more than three pounds of carbon dioxide is produced. We burn over one billion tons of coal every year. Four hundred and ninety-two coal-fired electric power plants were operating in 2007, with an average size of 667 megawatts and an average age of forty years. A single coal-fired power plant, rated at 500 mega-watts, produces about 3 million tons of carbon dioxide each year. If only 60 percent of the carbon dioxide from all of the 492 plants were captured, it would amount to about 2.4 million tons each day. The carbon dioxide, compressed into a liquid, would require about 20 million barrels each day for storage. To put this amount of liquid carbon dioxide into perspective, we consume about 20 million barrels of oil each day.

A number of demonstration plants are under construction using the technology and investment support of the DOE and other federal agencies.

Four projects are selected to provide some insight into the differences in coal plant operation. The purpose of each plant is to demonstrate performance and to investigate the potential economical production of electric energy from coal combustion. The four projects are:

- Integrated Gasification Combined-Cycle Electric Power Plant
- The Mountaineer Coal Plant
- The FutureGenProject
- The John W. Turk Jr. Power Plant

Integrated-Gasification Combined-Cycle Electric Power Plant

One type of coal-fired electric power plant is called an "integrated-gasification combined-cycle plant." Research and development for this type of plant is funded and supported by the DOE. It is called a "combined cycle" because the plant uses two types of turbines—a gas turbine and a steam turbine—to drive electric generators. The power plant operates by gasifying coal into a synthetic gas that is used to drive a hot gas turbine. The gas turbine is used to drive an electric generator. The hot gas exhaust from the gas turbine is used to generate steam. The steam is used to drive a steam turbine and electric generator to produce additional electric power. The technology for coal gasification, gas turbines, and steam turbines is well developed. The integration of these three technologies as applied to the design and operation of a coal-fired electric power plant is new and presents challenges. The DOE has reported that the combined-cycle plant is currently viewed as being too risky for private investors and may require large subsidies from federal, state, and local governments. Capital costs for the combined-cycle power plant are estimated at 20 to 47 percent higher than traditional coal plants.

In 2007, construction of an integrated-gasification combined-cycle facility was started near Orlando, Florida. Construction of the facility, scheduled to start operation in 2010, is a joint venture with the DOE, Southern Company, Kbr Inc., and the Orlando Utilities Commission. The facility is rated to produce 285 megawatts of electricity. The Clean Coal Power Initiative will turn coal into synthetic gas for generating electricity. The facility will significantly reduce emissions of sulfur dioxide, nitrogen oxides, and mercury when compared to existing pulverized coal plants. In addition, the facility will produce 20 to 25 percent less carbon dioxide emissions and use about 50 percent less water than existing pulverized coal-fired plants.

The appearance of a combined-cycle power plant more closely resembles a chemical plant than a conventional pulverized coal-fired power plant. Water becomes contaminated when it is used to clean the gas. As an example, the Great Plains Coal Gasification plant, located in Beulah, North Dakota, generated 5.32 million tons of wastewater in 1988, another 844,000 tons of contaminated "cooling tower blow-down" water, and 270,000 tons of gasifier ash, which will ultimately leach contaminants into the ground water where it is dumped. Ground water in the area has been contaminated with acids, sulfates, chlorine, arsenic, and selenium.

The Mountaineer Coal Plant

The American Electric Power Company, the largest consumer of coal in our country, is retrofitting its Mountaineer coal plant located in New Haven, West Virginia. The purpose of the plant modification is to demonstrate the capture of carbon emissions and store them underground. The plant, before modification, has been operating for about thirty years. The equipment needed for the removal of carbon emissions required a large volume of space. A four-story building that covers a land area somewhat larger than a football field is being used. In addition to the four-story building, an exhaust stack, which has a height of 150 feet, is part of the installation.

How does the process work?

- Liquid ammonia is used to absorb the carbon based emission gases. Then the carbon dioxide is separated from the ammonia.

- The gas is compressed into liquid form and injected more than a mile deep underground into a layer of porous sedimentary rock.

- A layer of impermeable shale, located above the sedimentary rock, prevents the liquid carbon dioxide from migrating to the earth's surface and should be permanently trapped.

An alternative for storing the liquid carbon dioxide would be to pump it into depleted oil or natural gas deposits, where it could be used to force remaining oil or gas deposits up to the earth's surface. Another possibility would be to pump the liquid carbon dioxide into coal deposits that cannot be mined. The carbon dioxide would force methane gas to other underground locations, and eventually the carbon dioxide would be absorbed into the coal.

Brad Linscott

The new equipment being installed at the New Haven, West Virginia, coal-fired electric plant will only capture 15 percent of the emissions generated by the plant. The new equipment will take the exhaust from the coal-fired plant and "bubble" it through a solution of chilled ammonia. The carbon dioxide will bond with the ammonia and be separated from other gases. Then the carbon dioxide will be separated from the ammonia and compressed for storage.

The equipment used to capture and store carbon dioxide requires large amounts of electric energy, thereby reducing the plants transmission of electric energy to its customers. The equipment requires 15 percent of the power plants energy output. Other types of emission-control equipment use as much as 30 percent of a power plants energy output. A utility company using carbon capture equipment may require purchasing electric power from other sources due to the shortfall. Officials from the American Electric and Power Company estimate the cost of carbon capture for a modest-sized plant coal plant rated at 235 megawatts starts at $700 million (including the cost to buy additional power to make up for the power lost).

The FutureGen Project

A new project, called FutureGen, will be led by FutureGen Industrial Alliance, Inc.[24] This company is a non-profit industrial consortium representing the coal and power industries. FutureGen is an initiative to build a first-of-its-kind, coal-fired, near-zero-emissions power plant. The plant, to be located in Matoon, Illinois, will establish the technical and economic feasibility of producing electricity from coal, while capturing and sequestering the carbon dioxide generated in the process. The project will be a government/industry partnership to pursue the design, construction, and "technically cutting-edge" power plant. The intent of the project is to eliminate environmental concerns associated with coal utilization.

The project will employ coal gasification technology integrated with combined-cycle electricity generation and the sequestration of carbon dioxide emissions. The project will be supported by results from the DOE's ongoing coal research program. The DOE research results will provide the technology to support the project. Under the terms of the DOE/FutureGen partnership, the planned activities by early 2010 include: preliminary design, completion of site-specific preliminary design and cost estimates, expansion of sponsors for the project, and development of a complete funding plan.

The DOE's total anticipated financial contribution for the project is $1.07 billion, $1.0 billion of which comes from the American Reinvestment and Recovery Act 2009 funds allocated for carbon capture and storage research. The FutureGen Alliance's total anticipated financial contribution is $400 million to $600 million based on a goal of twenty member companies each contributing a total of $20 million to $30 million over a four-to-six-year period. The alliance, with support from the DOE, will pursue options to raise additional non-federal funds needed to build and operate the facility, including options for capturing the value of the facility that will remain after conclusion of the research project.

The John W. Turk Jr. Power Plant

As an alternative to carbon dioxide collection and storage, authorities in Arkansas in late 2008 gave the go-ahead for a 600-megawatt coal-powered facility proposed by the American Electric Power Company.

Developers claim the new John W. Turk Jr. Power plant, located in Arkansas, will be operated by the American Electric Power Company's subsidiary, Southwestern Electric Power Company. The plant will use low-sulfur coal and "state-of-the-art" emission control systems. The emission control technology will enable the Turk plant to meet emission limits that are among the most stringent ever imposed on a coal-fired facility and will make it one of the cleanest plants ever built.

The Turk Plant project is the only one in the country using the "ultra-supercritical" generation process, which burns coal as a generating fuel at much higher than typical temperatures, thereby burning more of the fuel and allowing less carbon to escape. The plant has been designed to incorporate additional technology for carbon capture once it becomes economically available, but since carbon dioxide is not considered a "regulated pollutant" under the definition of the Federal Clean Air Act, its capture and storage is not required.

Despite the support for the Turk coal plant in Arkansas, other states are taking a tougher line on plans for the construction of new coal-fired electric power plants. Former California governor Arnold Schwarzenegger recently proposed a policy that would impose stringent carbon emission standards on new coal-fired power plants and would effectively ban plants built without carbon capture and storage technology.

The storage of liquid carbon dioxide has its own challenges. Pumping the liquid into the ground is required for final storage. To safely store the liquid underground requires a site selection process because most of the earth is too porous to seal off the underground liquid. At the coal-fired plant near New Haven, West Virginia, the carbon dioxide will go into a saline-bearing rock formation called an *aquifer*. In other parts of the country, storage can be established below geological caps.

Transporting Carbon Dioxide

Many existing coal plants, if converted to control carbon emissions, or new coal plants having carbon emission control will have to be connected to pipeline networks. The pipelines are needed to carry pressurized gaseous carbon dioxide to areas more suitable for underground storage. Depending on the length of the pipelines, pumping stations periodically located along the pipeline are required to maintain the flow of carbon dioxide. When the gas arrives at the storage site, the gas will be compressed to a pressure high enough to convert the gas into a liquid. The temperature of the carbon dioxide gas must be reduced below 87.8 degrees Fahrenheit. The temperature at 87.8 degrees is called the *critical temperature* for carbon dioxide gas. For a gas temperature above the critical temperature, no amount of pressure, however great, will cause the gas to liquefy. A pressure of 72.9 atmospheres (or 1,057 pounds per square inch, as measured with a pressure gauge) is required to convert the carbon dioxide gas, at a temperature below 87.8 degrees Fahrenheit, into a liquid. It has been estimated that over the next forty to fifty years, if proposed new carbon emission controls are enforced, billions of tons of liquid carbon dioxide will be injected underground. In addition, these billions of tons of carbon dioxide must be stored without leaking for thousands of years.

In order for the liquid to remain liquid, the temperature must remain below 87.8 degrees. For deep penetrations into the earth, the earth temperature increases with depth. The temperature of the liquid carbon dioxide may increase below ground. If the temperature exceeds the critical temperature, the liquid will begin to turn into a gas. For a gas temperature above the critical temperature, no amount of pressure will cause the gas back into a liquid state. If the underground liquid turns into a gas, it will result in quite an explosion with an effect much like an earthquake.

The question has been raised as to why many profit-making organizations are financially contributing to and are continuing to lobby Congress to support clean coal technology. Why are our electricity producers, the automobile, oil, and coal industries, and our railway corporations favoring clean coal technology, including carbon dioxide capture and storage?

- Is it because electricity producers can forget about renewable energy if they can begin to capture and store carbon dioxide on a large scale? They can stop worrying about unfamiliar renewable technologies like solar, wind, geothermal, and tidal power.

- Do coal mining executives and the coal miner's unions strongly favor carbon dioxide capture and storage because without it they are likely to slowly go out of business? To avoid a total phase out of coal, the coal industry is desperately eager to demonstrate that carbon dioxide can be captured and stored below ground level.

- Do the railway corporations want coal-fired electric power plants to successfully capture carbon dioxide and store it? They may be in favor of carbon capture because 44 percent of all rail freight, by weight, is coal.

- Are the automobile and oil corporations enthusiastic to have coal-fired power plants implement carbon dioxide capture and storage as part of each power plants operation? An 85 percent reduction of carbon dioxide emissions from coal-fired plants would make "space" for the automobile industry and the natural gas industry to continue to pollute the air we breathe.

The Story at Lake Kivu

Lake Kivu is located along the Democratic Republic of the Congo and the Rwanda border.[25] About 2 million people inhabit areas surrounding Lake Kivu. Scientists familiar with that lake region believe that a massive amount of carbon dioxide and methane is trapped below a depth of 1,000 feet under the lake. The gases are kept from rising to the surface of the lake due to the weight of the lake's water above. Sudden underwater landslides or other geologic activity have the potential to churn the lake enough to allow the gas to bubble to the surface of the lake. Carbon dioxide gas is heavier than air. The gas will hug the ground and can spread over large populated areas near the lake. If carbon dioxide gas leaks into the atmosphere from the water and spreads to surrounding land areas, the gas could asphyxiate many humans and animals.

Such a disaster occurred at Lake Nyos in Cameroon, West Africa in 1986. One thousand seven hundred people suffocated, many quietly in their sleep, due to high concentrations of carbon dioxide gas in the air. It is estimated that Lake Kivu has trapped 350 times the amount of gases as were trapped under Lake Nyos. A far larger number of people currently live in areas that surround Lake Kivu compared to the number of people that were living near Lake Nyos at the time of the disaster. The potential for death by asphyxiation in the populated areas near Lake Kivu is far greater than the disaster of 1986.

Is the plan to store liquid carbon dioxide underground feasible and can it be safely contained for hundred of years into the future? Or will the carbon dioxide stored underground slowly seep upward toward the earth's surface and affect our drinking water taken from our rivers and lakes? Remember the 1,700 people that died from carbon dioxide asphyxiation at Lake Nyos in Cameroon, West Africa. Will we be setting the stage for a man-made disaster similar to nature's disaster at Lake Nyos? Or are we likely to cause a potential disaster similar to the situation at Lake Kivu?

An alternative to carbon capture and storage, called "terrestrial sequestration," is being developed by federal agencies. Terrestrial sequestration involves the net removal of carbon dioxide from the atmosphere by plants and microorganisms that use carbon dioxide in their natural cycles. Terrestrial sequestration requires the development of technologies to quantify with a high degree of precision and reliability the amount of carbon stored in a given ecosystem. Program efforts in this area are focused on increasing carbon absorption by plant life. The program will evaluate plant life on mined lands, no-till agriculture, reforestation, rangeland improvement, recovery of wetlands, and river and creek bank restoration. These activities complement collaborative research with the US Department of Agriculture, the DOE Office of Science, the Environmental Protection Agency, and the US Department of the Interior.

Do we want more pipelines spanning our country for the purpose of transporting carbon dioxide from coal-fired plants to underground storage sites? The storage sites are likely to be located hundreds of miles away from most of our coal-fired plants. We have experienced problems with flawed pipelines, such as the Alaskan oil pipeline, that can cause environmental safety issues.

The pilot coal plants, with carbon control and underground carbon dioxide storage being built and put into operation, may provide us with a degree of confidence within the next five to ten years. Will the large companies that support underground storage try to convince us by advertising that carbon dioxide storage underground is completely safe based on the results of a few years of tests? Remember the promotional advertisements during the 1950s for filtered cigarettes? Those advertisements, as an example, tried to convince new and regular smokers that cigarettes were safe from the standpoint of human health, based on cigarette company tests.

Now you know why billions of our tax dollars are being used to support the development of carbon dioxide capture and storage. Do we as citizens, having common sense, want to continue spending our money to further develop this technology? There are serious questions that must be addressed before our country agrees to accept the risks of underground storage and pipeline transportation of carbon dioxide. Many human lives and our natural and financial resources are at stake if some of the billions of tons of carbon dioxide, stored underground, begin to escape from the ground into the atmosphere. If this activity is allowed to continue, we can expect in the future hazardous leaks from either underground or from above-ground pipelines transporting carbon dioxide under pressure.

The DOE together with the Fossil Energy Regional Carbon Sequestration Partnership program are beginning to implement large-scale carbon dioxide storage tests in locations throughout our country.[26] What if some of these storage sites encounter an unexpected malfunction and large amounts of carbon dioxide gas begins to enter the atmosphere, putting human life and property in jeopardy? Our government, using our tax dollars, is liable to correct such an occurrence. If we allow this activity to continue and allow the number of underground storage sites to increase, we will be setting the taxpayer up for another government "bailout" in the future.

The Department of Energy's FY 2010 budget requested $5.8 billion to protect human health and safety to clean up hazardous, radioactive legacy waste from the Manhattan Project and the Cold War. This funding, along with American Reinvestment and Recovery Act 2009 money, in the amount of $6 billion, will allow the program to continue to accelerate cleaning up and closing sites. Will we face a similar situation if in ten years carbon dioxide is expelled into the atmosphere at test sites located throughout our country? Do we really want to allow large-scale carbon dioxide storage tests in locations throughout our country?

Safety Issues with Underground Coal Mining

More than 100,000 coal miners have been killed in accidents in our country since 1900. The number of deaths has fallen sharply in recent decades according to the Mine Safety and Health Administration. A coal mine disaster in April 2010 claimed the lives of twenty-nine coal miners from Montcoal, West Virginia. Federal officials indicated that the accident was the worst, in terms of the number of deaths, in twenty-five years. The underground mine explosion occurred at the Upper Big Branch mine located thirty miles south of Charleston, West Virginia.

A similar explosion occurred in 2006 at the Sago Mine in West Virginia that killed twelve underground miners. In December, 1984, twenty-seven underground miners died in a fire at the Wilberg Mine located in Orangeville, Utah.

Unfortunately, more of our coal miners will be killed in accidents that are likely to occur in the future. Many coal miners will be, or have been, stricken with black lung disease, as a result of their work in underground coal mines. Should our country continue to increase underground coal operations or phase out coal production for coal-fired power plants in favor of environmentally cleaner energy sources?

Natural Gas Electric Power Plants

Natural gas-fired electric generation plants provide about 20 percent of our energy needs. In the states of California and Texas, natural gas-fired generation provides nearly 50 percent of the electricity consumed. Natural gas is clean-burning and efficient and is a popular fuel for the generation of electricity. In the 1970s and 80s, the fuel choice of most electric utility companies was coal or nuclear power. But as environmental concerns grew and technology advanced, natural gas became the fuel of choice. During the last ten years, over 90 percent of the new electric capacity built in our country has been natural-gas-fired generation. Natural gas electric generation plants cost less to build and are generally more acceptable to the public when compared to coal-fired or nuclear power plants. It is usually easier to obtain local, state, and federal agreement on construction sites and the associated permits needed for new gas-fired power plants than it is for coal-fired or nuclear plants.

While coal is the cheapest fossil fuel for generating electricity, it is also the dirtiest, releasing the highest levels of pollutants into the air. The electric generation industry, in fact, has been one of the most polluting industries in our country. Regulations surrounding the emissions of power plants have forced these electric generating plants to come up with new methods for generating power while reducing environmental damage. New technology has allowed natural gas to play an increased role in the generation of electricity.

Steam Boiler Plants

Natural gas can be used to generate electricity in a variety of ways. One method is to burn the gas to heat water in a boiler that produces steam.[27] The steam is used to drive a turbine that is mechanically connected to an electric generator. Using steam to produce electric power is most often used by large coal-fired or nuclear power plants. The plants using the basic steam generation operation have relatively low efficiencies, about 33 percent, compared to plants using newer technologies. The efficiency is measured by comparing the heat available in the fuel to the steam generated in British Thermal Units (BTUs) and kilowatts produced by the electric generator.

Gas Turbines and Internal Combustion Engines

Gas turbines and internal combustion engines are also used to generate electricity. For gas turbines, instead of heating steam to drive a turbine, hot gases from the combustion of natural gas are used to drive the turbine. The natural gas is injected into a container called a *combustor*. The combustor is designed to ignite and burn the natural gas and air mixture to a very high temperature and pressure. The high-pressure hot gas mixture is ducted to flow from the combustor to the turbine, causing the turbine to rotate. The electric power plant gas turbine is connected to an electric generator. Gas turbine and combustion engine plants are traditionally used to meet peak-load demands. The gas-fired plants have the capability to start operation quickly compared to the water boiler plants. Gas turbines are also used to propel the so-called prop-jet airplanes. Gas-fired plants have increased in popularity due to advances in the technology and the availability of natural gas. However, these plants operate at a slightly lower efficiency—less than 33 percent—than the steam-driven power plants.

Combined-Cycle Power Plants

Many of the new natural-gas-fired electric power plants are called *combined-cycle units*. This technology is called a "combined-cycle" because it requires a gas turbine and a steam turbine. The flow of hot pressurized gas released from the combustion of natural gas is used to turn a turbine-generator combination. In the combined-cycle plants, the gas-turbine exhaust gas, called "waste heat," is directed to a steam generator. The steam is used to drive a steam turbine-generator combination. Because more heat is extracted from the natural gas, combined-cycle plants are more efficient than steam units or gas turbines alone. The combined-cycle gas-fired electric plants achieve thermal efficiencies between 50 to 60 percent, a significant increase compared to 33 percent for steam plants.

Advocates for natural gas as an energy source agree that there is widespread support for moving the economy into a new era based on renewable and clean sources of energy. But the reality for renewable energy is that we will be using hydrocarbons for decades to come. The reason being that the technology and infrastructure required for renewable energy sources are both being developed. Therefore, it only makes sense to promote natural gas as a "bridge fuel" to achieve these environmental goals.

The current nonrenewable options for replacing an anticipated loss of base-load generating capacity are coal-fired thermal and combined-cycle gas-combustion turbines. It is anticipated that the demand and price for cleaner burning natural gas will escalate substantially during the next twenty-five years. An increase in the amount of imported gas may be needed when new gas-fired power plants become operational. Increasing the amount of imported natural gas would further compromise our energy security beyond the continuation of oil imports that are needed to meet our transportation commitments. Local, regional, and global environmental impacts associated with increased coal use will most likely require a transition to clean-coal generation that requires the sequestration of carbon dioxide. The costs and uncertainties associated with such a transition are daunting.

Some electric utility executives say they have little choice but to see plants fired by natural gas as the only kind that can be constructed quickly and can supply reliable power both day and night. The price of natural gas tripled in the late 1990s and early in this decade. Part of the reason for the spike in natural gas prices was because a large number of electric plants started to use natural gas. During that period, some of the natural-gas-fired power plants were not economical to operate compared to coal-fired plants. Many plans for new coal-fired plants are now being canceled. With the demand for electricity rising by about 1 percent each year, the prospect is that utilities will be forced to build and use a new generation of gas-fired plants regardless of the operating cost, and consumers will bear the burden of higher utility rates. In the meantime, some utility companies have decided to wait for a clear-global-warming policy to emerge from one of the administrations in Washington.

The Florida Light and Power Company provides a good example of a company that has shifted from coal to natural gas as an energy source. The company has about 4.5 million customers. It is adding about 85,000 new customers each year, and demand for electricity from their existing customers is increasing. In 2007, the Florida Public Service Commission stopped the company's plans to build a large coal-fired electric plant. The plant was planned to be built near the Everglades National Park. The new coal-fired plant was to start operation in 2014. When plans for the new plant were canceled, the utility company analyzed the possibility of replacing the coal-fired plant with various types of renewable energy resources including solar power. Their analysis concluded that renewable sources of energy could only replace a small fraction of the electric power needed to replace the planned-for coal-fired plant. As a result, the utility company decided to accelerate construction on a long-planned addition to an existing natural-gas-fired electric plant.

Do We Have Enough Natural Gas?

We have continued to hear that our energy policy should be directed to reduce our dependence on foreign oil. How about our dependence on imported natural gas? During the first quarter of 2009, January through March, we imported, from foreign countries, a total of 1,021 Bcf (billion cubic feet) of natural gas.[28] Most of our imported natural gas comes from Canada. We also export natural gas to foreign countries. During the first quarter of 2009 we exported 331 Bcf of natural gas. Most of our exports are sent to Canada. The difference between what we import and the amount we export amounts to a net importation of 690 Bcf of natural gas, just for the first three months of 2009. During the whole year of 2009, about four times the first quarter amount or a total of about 2,760 Bcf was imported.

The amount of imported natural gas is estimated at 25 percent of the total consumed in 2009.[29]

Residential customers consumed 4,865 Bcf in 2008. Commercial customers used 3,121 Bcf, and industrial users consumed 6,618 Bcf. A number of road vehicles using natural gas rather than gasoline used 30 Bcf. And our gas-fired electric power plants used 6,660 Bcf. The remainder of natural gas is consumed under the heading of "Lease and Plant Fuel," which amounts to 1,284 Bcf for 2008. If we compare the total one-year net imports (2760 Bcf) with the amount consumed by residential users (4865 Bcf), we imported about 57 percent of the natural gas used by residential customers.

Drilling for Natural Gas

In our country wells are drilled into the ground to remove natural gas found below. After the natural gas is extracted, it is treated at gas plants to remove impurities such as hydrogen sulfide, helium, carbon dioxide, hydrocarbons, and water. Pipelines then transport the natural gas from the gas plants to electric generating plants.

The most fertile locations for new gas wells tend to be found in backyards in Arkansas, state forests in Pennsylvania, and ranches in Colorado.[30] These new fields are expanding as a result of the precipitous decline in offshore fields and other onshore reserves that are losing pressure with age. Companies have been forced to move into less-prolific territory merely to keep production stable, let alone meet growing demand. It is expected that these will supply 80 percent or more of the increase in domestic production anticipated through 2030, as reported by the US Energy Information Administration.

Gas Wells in Pennsylvania and New York

In the states of New York and Pennsylvania, a few thousand feet underground, lay the "Marcellus Shale" beds. These shale beds are impregnated with natural gas. It is estimated that the Marcellus reserve contains at least 350 trillion cubic feet of natural gas. This quantity is enough to supply the current US demand for between ten and fifteen years. In order to release the natural gas and extract the gas for storage, the underground shale beds must be broken up or fractured. The process to capture these gas reserves is called hydro-fracturing, or "fracking." In order to break up the underground shale beds, the fracking process first involves drilling a hole about one mile deep into the shale bed. Next a large amount of potable water is mixed with toxic chemicals. This mixture is forced deep into the shale bed, under pressure, causing the shale bed to break up and thereby release the entrapped natural gas.

This drilling activity is in the early stages of development. Some energy companies claim that the injected fluids are well below the level of drinking water aquifers. And they further claim that the chemicals added to the water are so diluted that the mixture poses no threat to health. Some Pennsylvania environmental officials advise that the process of fracking will result in some environmental damage, including possible contamination of water supplies.[31] Pennsylvania officials admit that some of the chemicals that are added to the water used for fracking could be dangerous to human health. Some officials believed that the monetary value of the gas underlying Pennsylvania and parts of surrounding states outweigh the environmental damage. In other words, the money earned from the sale of natural gas by the gas companies and the associated job opportunities outweighs any environmental damage and risk of jeopardizing human health that potentially can occur.

Many of the now-producing gas wells are located in northeast Pennsylvania, near the township of Dimock. There have been some reports that residents of Dimock have experienced polluted drinking water. They say their water has caused sickness and, at times, has become discolored and foul smelling since drilling started.

Many wastewater treatment plants located near areas where gas drilling is taking place are eager to gain the revenue from accepting gas drilling wastewater, called "fracking fluid." Some Pennsylvania wastewater treatment plants that have accepted fracking fluid for processing have created problems for the downstream drinking water plants. In Pennsylvania, several water treatment plants located downstream from the wastewater plants began to measure abnormally high concentrations of solid materials, sodium, and chlorides. The measurements were taken after the wastewater plants began to process fracking fluids and compared to measurements taken before. One wastewater plant reportedly discovered that some of the waste was radioactive. A wastewater treatment plant located in Bath, New York, began accepting fracking waste.[32] The state Department of

Environmental Conservation recently found a sharp increase in dissolved particles when downstream discharge water was tested.

Gas Wells in Ohio

Newly drilled natural gas wells in Ohio have improved the state's economy by opening new jobs and providing new business for local companies connected with gas drilling operations. On the other hand, some wells have scarred the landscape and created significant wounds to the environment. Current law in Ohio does not allow local control by municipalities to protect citizens' safety or to prevent environmental destruction. In affected parts of Ohio where gas drilling has been carried out, large wooded areas have been eliminated, water runoff has occurred, and safety issues have been ignored.

Some Ohio municipalities have passed ordinances to protect its citizens from abuses caused by some of the natural gas drilling operations. However, the city ordinances cannot, because of state and federal law, be binding on the part of the gas companies. If a gas company adheres to a city ordinance, they do so voluntarily. In most cases, the gas drilling companies do not abide by local municipality ordinances that are designed to protect property and its citizens. Some Ohioans in suburban Cleveland, for example, were evacuated from their homes due to a gas-well leak. Some families had their well water contaminated with gas. They now have to drink bottled water delivered to their home once a week. Some website video demonstrations show a cigarette lighter being ignited near a stream of water running from a kitchen faucet. Almost instantly, the natural gas mixed with the water ignites and the kitchen faucet spews flames like an old-fashioned blowtorch. Some affected Ohioans see huge, newly installed storage tanks and a pumping rig while looking out their window.

As another example, a suburban Cleveland resident bought a home with property adjoining a wooded area. The wooded area belonged to another property owner. Recently, before leaving for work one morning, the owner heard the noise of tree cutting machines and limb shredders from the wooded area behind the home. The property was being cleared to allow for the drilling of four oil wells. The owner now estimates that the value of his home has decreased by $50,000. This suburb has no effective legislation to control the installation of the gas wells and the potential adverse effects on nearby residential properties. The gas that sometimes leaks into the air as a result of drilling for oil is odorless, thereby presenting a safety risk to nearby residents and their homes.

Gas and oil wells do produce a profit for Ohio. But they also can create safety risks and environmental hazards and decrease property values, and the activity can damage roads that are not constructed to carry heavy drilling equipment and large heavily loaded trucks. In Ohio these examples indicate that the Ohio Gas Association and the Ohio Department of Natural Resources are not working together to serve the best interest of property owners located adjacent to or near newly developed oil and gas drilling sites.

Finding a way to transport natural gas from new gas wells to consumers will likely be the next battle for producers. The Federal Energy Regulatory Commission approved the construction of one pipeline over 100 miles in length in 2004, three in 2005, six in 2006, and nine in 2007. One proposed pipeline will stretch from Colorado to New Jersey. But to fill those pipelines, producers must convince thousands of landowners to allow smaller lines to crisscross their property, connecting individual wells and gathering stations. One complaint from the industry is that there's nobody in power to make a final decision when there are conflicting demands.

Nuclear Electric Power Plants

The nuclear electric power plants that are now in operation use a nuclear fission reactor to generate heat. The heat is transferred to a supply of water causing the water to evaporate into steam under pressure. A flow of high-pressure steam is directed to a steam turbine connected to an electric generator. As an example, the core of a 1,000-megawatt nuclear reactor contains about seventy-five tons of enriched uranium. The water coolant is pumped through the reactor and carries away the heat produced from the splitting of atoms called nuclear fission. The resulting superheated steam is used to drive a steam turbine that is mechanically connected to an electric generator. A plant rated at 1,000 megawatts produces about 8 billion kilowatt-hours (kWh) of electricity per year. To maintain efficient nuclear reactor performance, one third of the spent fuel is removed every year and replaced with fresh fuel.

Nuclear power plants generate about 20 percent of the electricity produced in our country. Most recent electric generation capacity additions and projected future additions are primarily fueled by natural gas. Despite the excellent performance of our nuclear plants and decisions by power plant owners to seek license renewals and power plant updates, no new plant has been ordered in our country for more than twenty-five years.

In 2004, one hundred and four nuclear power plants were on line. Total nuclear-electric production has increased from about 630,000 gigawatt-hours in 1994 to 790,000 gigawatt-hours in 2004. These plants have accrued a total increase in energy production of 26 percent in ten years. This was accomplished without bringing a single new nuclear unit of electrical generating capacity on line. It is noteworthy to add that during the period of 1994 to 2004, about 7,000 megawatts of peak capacity was either retired or idled. Technological improvements have rapidly increased the efficiency and "load factor" of all nuclear power plants. The percentage of the year's time they spend producing at or near full capacity has increased from 70 percent to over 90 percent, far higher than power from other fuels.

New base-load nuclear generating capacity is required to enhance our energy supply diversity and energy security. The Nuclear Power 2010 program (NP2010), unveiled by the Secretary of Energy on February 14, 2002, is a joint government/industry cost-shared effort. Its purpose is to identify sites for new nuclear power plants, to develop and bring to market advanced nuclear plant technologies, evaluate the business case for building new nuclear power plants, and demonstrate untested regulatory processes.[33] Accomplishing these program objectives paves the way for a government and an industry decision to build new, advanced light/water reactor nuclear plants in our country that would begin operation before 2020. The DOE is actively engaged with the industry to address the issues affecting future expansion of nuclear generation. Part of the Nuclear Power 2010 program was focused on reducing the technical, regulatory, and institutional barriers to deployment of new nuclear power plants. The DOE successfully completed that part of the Nuclear Power 2010 program in FY 2010 as reported in the Department of Energy FY 2011 Congressional Budget Request.

The FY 2009 appropriation for the program was $177 million, and $220 million was appropriated for FY 2010.

The remainder of the Nuclear Power 2010 program manages the design effort, ordering long-lead items of equipment and planning for the construction and installation of thirty-four new nuclear reactor electric power plants.[34] Three early site permits have been issued. Two reactor designs have been certified, and four reactor designs are being reviewed. General Electric's Advanced Boiling Water Reactor design, rated at 1,356 MWe, has been certified. General Electric has submitted the design for review of an Economic Simplified Boiling Reactor rated at 1,560 MWe. The Westinghouse Advanced Passive Water Reactor design, which has twin units rated at 1,117 MWe each, has been certified. AREVA has submitted the design for the United States Evolutionary Pressurized Water Reactor rated at 1,600 MWe. Mitsubishi Heavy Industries submitted the design for the United States Advanced Pressurized Water Reactor rated at 1,600 MWe.

Twenty-one nuclear plant locations have been announced. Four nuclear plants will be located in Texas, two plants in Florida, and two plants will be located in South Carolina. One nuclear plant will be sited in each of the following states: Alabama, Georgia, Idaho, Illinois, Louisiana, Maryland, Michigan, Mississippi, Missouri, New York, North Carolina, Pennsylvania, and Virginia.

There is growing support in our country by corporate and government leaders for construction and operation of a large number of nuclear electric power plants.

In 2008, Assistant Secretary of Energy Spurgeon suggested that our rising energy demands, our security, our prosperity, and our environment all require reducing our dependence on fossil fuels that emit greenhouse gases. He asserted that no serious person can look at the challenge of maintaining our national security, reducing greenhouse gases, and addressing climate change and not come to the conclusion that nuclear power has to play a significant and growing role. He concluded that to foster that growing role, our nuclear energy policy itself must take on a more significant role to be technologically robust, economically sound, and publicly acceptable for decades to come. An economically sound policy is one that reduces financial barriers to deployment of new nuclear power. Currently that means loan guarantees, risk insurance, and production tax credits.

In 2008, the CEO of Progress Energy Florida stated that nuclear plants are the most expensive to build, but customers will pay less in the long run because the cost of generating electricity from a nuclear plant is far below the cost of making it from coal, natural gas, wind, or solar. According to industry estimates, the cost of generating electricity from a nuclear plant is about 0.4 cents a kilowatt-hour, 4.2 cents from a coal plant, and 7.0 cents from a natural gas plant. Over its lifetime, the nuclear power plant will have the lowest fuel cost and it will have the lowest environmental impact. Nuclear energy reduces our dependence on foreign fuels, it provides long-term cost stability for customers in that we're not as dependent on volatile and expensive natural gas and oil, and it doesn't produce any greenhouse gases.

Although initial construction costs are high, our nuclear electric plants have proven to produce electricity at a cost that is an order of magnitude below that of any other type of electric plant in operation.

Nuclear power plants currently operating in our country benefit from federal subsidies. The potential cost of damages that might result from an accident at a plant site may be too large for the insurance industry to cover. The federal government has pledged to act as the "insurer of last resort" above a certain cost level.

The DOE has identified three major missions for Generation IV systems:

- Actinide management
- Electricity generation
- Hydrogen (or other non-electricity products).

Actinide Management

Actinide management has significant benefits for our society. It can initiate the consumption of nuclear waste in the mid term and provide assurance of nuclear fuel availability in the long term. Actinide management addresses the disposition of used nuclear fuel and high-level nuclear waste. The so-called transuranic actinides are the radioactive elements: neptunium, plutonium, americium, and curium. The currently operating nuclear electric plants that supply about 20 percent of the total electricity in our country generate these transuranics. Large inventories of these radioactive elements have been accumulating since the 1960s. More than twenty metric tons of transuranic elements are produced annually. Complete recycle of the transuranics produce energy for electricity generation or for other purposes. In the process of producing energy, a great proportion of the long-lived radio-toxic constituents are destroyed, where otherwise they would require isolation in a geologic-repository.

The potential for nuclear energy to play an increasing role in energy production has increased significantly. Since 2003, the number of our nuclear plants filing for extensions of their operating licenses from forty to sixty years has increased. The DOE expects that the licenses of over 75 percent of all reactors may eventually be extended. The industry is investing heavily in existing nuclear plants for capacity increases that could provide the equivalent of several new plants. The Energy Policy Act of 2005 included incentives for new nuclear plants in order to reduce the financial risk, thereby increasing industry interest.

Another trend to note is the recent increase in uranium prices. The ongoing United States/Russian program on downblending weapons-grade uranium has provided fuel for domestic commercial reactors. There is a sufficient amount of fuel to last for several years, and it will partially help to meet our uranium needs. However, domestic license extensions, new plants being built by other nations, and the potential of new domestic plants all contribute to increased uranium demand projections. In response, spot prices have increased and the number of applications for new uranium mining is now rising. The DOE has proposed a Global Nuclear Energy Partnership (GNEP) that would seek to develop worldwide consensus on enabling expanded use of economical, carbon-free, closed nuclear fuel cycle that enhances energy security while improving proliferation risk management. It would achieve its goal by having nations with secure, advanced nuclear capabilities provide fuel services, fresh fuel, and recovery of used fuel to other nations who agree to employ nuclear reactors for power generation purposes only.

The Global Nuclear Energy Partnership will:

- improve the utilization of geologic waste repositories by recycling transuranic elements. This effort will help to assure that only one geologic waste repository would meet our need for the remainder of this century.

- enable the recovery of the energy content in the transuranic elements and separate excess uranium for possible future use. The current once-through fuel cycle used in this country only extracts about 1 percent of the energy content in the original uranium ore. The remaining energy value resides in used nuclear fuel, including the transuranic elements and the uranium that is depleted during the uranium enrichment process.

- design modern safeguards directly into the planning and building of new nuclear energy systems and fuel-cycle facilities. These new safeguards will allow for more effective and efficient monitoring and verification of nuclear material.

Electricity Generation

The Department of Energy has been evaluating concepts for three types of fast reactors systems. The three types are the sodium-cooled fast reactor, the lead-cooled fast reactor, and the gas-cooled fast reactor system. The DOE estimates that deployment of the lead-cooled fast reactor would be between five and ten years away, while deployment of the gas-cooled fast reactor would be about twenty years away. The DOE believes the challenges for sodium-cooled fast reactor technologies are well understood.

All three fast reactor systems perform very well on objectives related to waste management. These reactors can destroy transuranic materials that otherwise may present proliferation concerns. They perform very well on their ability to extract the maximum energy from uranium resources. Each reactor system has the ability to support the sustainability of nuclear energy for the very long term. The DOE has evaluated the economics of the three systems. The power density of the reactor core drives the size of many components and structures.

The Sodium-Cooled Fast Reactor

The DOE decided, for near-term deployment of fast reactor technology, the option with the most viable technical maturity is the sodium-cooled fast reactor. The sodium-cooled fast reactor is best suited to produce electricity.

The sodium-cooled fast reactor design provides a higher-power density of the reactor core and allows for a more compact system when compared to the other two systems being considered. From the standpoint of safety, sodium interacts energetically with water, but the possibility of such an interaction can be controlled. A number of sodium fast-test reactors have been operated both domestically and internationally for long periods of time with no coolant interaction problems.

Further development of the sodium-cooled fast reactor relies on technologies already developed and demonstrated. Sodium-cooled reactors and associated fuel cycles have successfully been built and operated in worldwide fast reactor programs. Approximately 300 reactor years of operating experience has been accumulated for sodium-cooled reactors. This experience includes 200 years on small test reactors and 100 years on larger demonstration or prototype reactors.

In our country, sodium-cooled reactor technology was demonstrated in the 20-megawatt electric experimental breeder reactor. This reactor operated from 1963 to 1994. The DOE Fast Flux Test Facility, a 400-megawatt thermal facility, was completed in 1980. The facility operated successfully for ten years and was used to perform sodium-cooled fast reactor materials, fuels, and component testing. Sodium-cooled fast reactor experience also extends to the commercial sector with the operation of Detroit Edison's FERMI 1 plant from 1963 to 1972.

Significant sodium-cooled fast reactor research and development programs have been conducted in Russia, Japan, France, India, and the United Kingdom. The only current fast reactor for electric generation is the Russian BN-600. It has reliably operated since 1980 with a 75 percent capacity factor. Fast reactors currently undergoing testing are located in France, Japan, and Russia. The most modern fast reactor, operating in Japan, is the 280-MWe plant completed in 1990. Sodium-cooled fast-reactor technology programs have recently started in Korea and China.

Early in 2008, our country, France, and Japan signed an agreement to expand their cooperation on the development of sodium-cooled fast reactor technology. The agreement relates to their use of advanced reprocessing and fast reactor technologies and seeks to avoid duplication of effort by the three countries. These ongoing international R&D activities may help improve the performance for both energy and fuel cycle applications.

Recent cost studies estimate that the capital cost of current fast reactor designs may be about 26 percent higher than the cost to build a conventional light water reactor. The light water reactor design is currently used for the nuclear power plants that are operating here. Much of the difference in cost is because cost reductions have been achieved as a result of the actual experience of building hundreds of commercial nuclear power plants worldwide with the light water reactor.

There is no equivalent experience with fast reactor designs. As a result, the DOE plans to assess a number of potentially cost-saving improvements in the demonstration sodium-cooled fast reactor facility. Some of the potential improvements include:

- materials selected and qualified for a sixty-year life.

- seismic isolation system for the reactor to demonstrate feasibility and potential benefits.

- advanced digital instrumentation and control.

- simplify the fuel-handling system by development and demonstration of a specialized in-vessel handling machine.

- improve the operations and maintenance technology to include possible in-service inspection and repair, and develop remote handling and sensor technology as related to using sodium as a coolant.

Considerable effort is required to update our country's infrastructure for sodium fast reactors. Much of the required infrastructure has atrophied since the last domestic sodium-cooled fast reactor was shut down in the early 1990s. Key infrastructure needs include:

- Reacquire the engineering and knowledge base to safely fill, drain, and operate flowing sodium systems for cleanup and purification of the sodium coolant, chemistry control, heating and cooling systems, and instrumentation and control. This technology was well established in our country with liquid metal (including sodium) reactor programs in the 1990s and needs to be reestablished to effectively build domestic sodium systems.

- There are no domestic industrial fabrication and testing capabilities for metal-cooled fast-reactor components.

The DOE facility previously used to conduct research and development in liquid metal applications and for testing large sodium components in a prototype environment has been decommissioned (along with most of our fast reactor development facilities) and is no longer available for use.

- Facilities previously used for making both metal and ceramic oxide fast reactor fuels are currently operational but will require installation of new equipment.

- Many of the existing computer design codes for reactor design were developed twenty years ago. Computer code modifications are needed to more accurately model the physics phenomena that occur in a modern reactor.

- Fast reactor safety analysis tools developed in our country reflect the current standards and are utilized in all the major international fast reactor programs. As with the reactor computer design codes, improvements could provide a more accurate analysis by using modern simulation methods.

- The international licensing and regulation standards for fast reactors are severely outdated since the last fast reactor was built in 1990. The regulatory resources and competency to review fast reactor safety needs to be reestablished.

The new reactor will help to reestablish the infrastructure needed to deploy commercial-scale fast reactors and begin building the experience base of reactor operations and the pool of trained personnel. The Department of Energy estimates design and construction of the reactor facility could be completed, with initial operation, by 2020.

Industry in our country was an active participant in the domestic research and development of sodium-cooled fast reactor technology during previous DOE programs. The success of the new reactor program will rely on the early and extensive involvement of industrial partners. The DOE expects that the industrial partners will provide a significant amount of cost-sharing for this activity. The direct customers of this new technology are the energy companies who own and operate the reactors and fuel-cycle facilities. Additional customers include the vendors and architect-engineers who must build the new facility, with the ultimate beneficiary being the American public. The DOE plans to continue their partnership with our universities and national laboratories, where the focus of research will be more applied. The DOE continues to increase international collaboration to support the goals of the Global Nuclear Energy Partnership.

Hydrogen Production

The third part of the DOE Generation IV mission is to develop the nuclear technology for the purpose of producing hydrogen or other non-electricity products. One reason for the effort is to demonstrate that nuclear plants can produce hydrogen gas more economically than we are now able to produce it. Because of the importance of the Generation IV mission to our future economy, a detailed discussion on hydrogen production is presented later.

The activities carried on by the DOE under the heading of Generation IV Nuclear Energy Systems program will be called "Reactor Concepts RD&D" starting in FY 2011. Carry-on activity under the new heading will include the Next Generation Nuclear Plant project, R&D on Generation IV, and other advanced nuclear reactor concepts. The program will conduct R&D to support extending the life of our nuclear electric power plants in operation and work on Small Modular Reactors. The DOE has requested $195 million to support this activity during FY 2011.

Transmitting Electricity

The Electric Grid

Electric power is essential to our modern society. Economic prosperity, national security, and public health and safety cannot be achieved without it. Communities that lack electric power, even for short periods, have trouble meeting basic needs for food, shelter, water, and law and order.

In 1940, 10 percent of energy consumption in America was used to produce electricity. Energy consumption increased to 25 percent by 1970. Today, 40 percent of our energy consumption is used to produce electricity. Electricity has the unique ability to convey both energy and information. The increased availability of electricity has spawned an increase in the number of new products, services, and applications in factories, offices, homes, universities, and communities. The electric industry is one of the largest and most capital-intensive sectors of our economy. The total asset value of the electric industry is estimated to exceed $800 billion, with about 60 percent of this amount invested in electric power plants, 30 percent invested in electric distribution facilities, and 10 percent invested in electric transmission facilities.

Annual electric revenues, the nation's "electric bill," are about $247 billion and are paid for by America's 131 million electricity customers. The customers are comprised of nearly every business and household in our country. The average price paid for our electricity is about seven cents per kilowatt-hour.

There are more than 3,100 electric utility companies in America. About 200 companies of the 3,100 are stockholder-owned utilities that provide power to about 73 percent of their customers. About 2,000 companies of the 3,100 are public utility companies that are run by state and local government agencies serving about 15 percent of their customers. A little over 900 electric cooperatives provide electricity to about 12 percent of their customers. In addition, there are nearly 2,100 non-utility power producers in the United States. These power producers include both independent power companies and customer-owned facilities for energy production and distribution.

The bulk power system, or electric grid, consists of three independent networks: Eastern Interconnection, Western Interconnection, and the Texas Interconnection. These networks incorporate international connections with Canada and Mexico. Overall reliability, planning, and coordination are provided by the North American Electric Reliability Council, a voluntary organization formed in 1968 in response to the Northeast blackout of 1965.

Electric Generation

With the over 3,000 electric utility companies, America operates a fleet of about 10,000 power plants. Power plants are generally long-lived investments. The majority of the existing plants are thirty or more years old. America faces a significant need for new electric power generation because of the expected near-term retirement of many aging plants in the existing fleet. Growth of the information economy, economic growth, and the forecasted growth in electricity demand are added factors that support the need for new electric power generation. In this transition, local market conditions will dictate fuel and technology choices for investment decisions, capital markets will provide the financing, and federal and state policies will affect siting and permitting. It is an enormous challenge that will require a large commitment of technological, financial, and human resources in the years ahead.

Even with adequate electric generation, bottlenecks in the electric transmission system interfere with the reliable, efficient, and affordable delivery of electric power. America operates about 157,000 miles of high voltage (about 230,000 volts) electric transmission lines. While electricity demand increased by about 25 percent since 1990, construction of transmission facilities decreased by about 30 percent. In fact, annual investment in new transmission facilities has declined over the last twenty-five years. The result is grid congestion, which can mean higher electricity costs because customers cannot get access to lower-cost electricity supplies and because of higher line losses.

Transmission and distribution losses are related to how heavily the grid system is loaded. Nationwide transmission and distribution losses were about 5 percent in 1970 and grew to 9.5 percent in 2001, due to heavier utilization and more frequent congestion. Congested transmission paths, or bottlenecks, now affect many parts of the grid across the country. In addition, it is estimated that power outages and power quality disturbances, for example low voltage, cost the economy from $25 to $180 billion annually. These costs could soar if outages or disturbances become more frequent or longer in duration. There are also operational problems in maintaining voltage levels.

America's electric transmission problems are also affected by the new structure of the increasingly competitive bulk power market. Based on a sample of the nation's transmission grid, the number of transactions have been increasing substantially recently. For example, annual transactions on the Tennessee Valley Authority's transmission system numbered less than 20,000 in 1996. They exceeded 250,000 in 2010. The system is not designed to handle the current volume of transactions. Actions by transmission operators to curtail transactions for economic reasons and to maintain reliability (according to procedures developed by the North American Electric reliability Council) grew from about 300 in 1998 to over 1,000 in 2000.

There are significant impediments to solve the country's electric transmission problems. These include opposition and litigation against the construction of new facilities, uncertainty about the cost recovery for investors, confusion over whose responsibility it is to build, and jurisdiction and government agency overlap for siting and permitting. Competing land uses, especially in urban areas, leads to opposition and litigation against new construction facilities.

Electric Distribution

The "handoff" from electric transmission to electric distribution usually occurs at the substation. America's fleet of substations takes power from transmission-level high voltages and distributes it to hundreds of thousands of miles of lower voltage distribution lines. The distribution system is generally considered to begin at the substation and end at the customer's meter. Beyond the meter lies the customer's electric system, which consists of wires, equipment, and appliances. Many appliances have computerized controls and electronics that ultimately operate on direct current and their numbers are on the increase.

The distribution system supports retail electricity markets. State or local government agencies are heavily involved in the electric distribution business, regulating prices, and rates-of-return for shareholder-owned distribution utilities. Also in 2,000 localities across the country, state and local government agencies operate their own distribution utilities, as do over 900 rural electric cooperative utili-

Brad Linscott

ties. Virtually all of the distribution systems operate as franchise monopolies as established by state law.

The greatest challenge facing electric distribution is responding to rapidly changing customer needs for electricity. Increased use of information technologies, computers, and consumer electronics has lowered the tolerance for outages, fluctuations in voltages and frequency levels, and other power quality disturbances. In addition, rising interest in distributed generation and electric storage devices is adding new requirements for interconnection and safe operation of electric distribution systems.

A wide array of information technology is entering the market that could revolutionize the electric distribution business. For example, having the ability to monitor and influence each customer's usage in real time could enable distribution operators to better match supply with demand, thus boosting asset utilization, improving service quality, and lowering costs. More complete integration of distributed energy and demand-side management resources into the distribution system could enable customers to implement their own tailored solutions, thus boosting profitability and quality of life.

North America's world-class electric system is facing several serious challenges. Scant investment has been made in the country's high-voltage lines since 1980, even though electricity demand has doubled in that period with the proliferation of computers, air conditioners, and flat-screen televisions and the growth of the nation's population from 223 million to 305million.

Major questions exist about the ability of our electric system to continue providing citizens and businesses with relatively clean, reliable, and affordable energy services. The recent downturn in the economy masks areas of grid congestion in numerous locations across America. These bottlenecks could interfere with regional economic development. The "Information Economy" requires a reliable, secure, and affordable electric system to grow and prosper. Unless substantial amounts of capital are invested over the next twenty to thirty years, in a new generation, transmission and distribution facilities and service quality will degrade and electric costs will go up. These investments will involve new technologies that improve the existing system. The possibility of advances in the technology could revolutionize and change the current configuration of the electric grid.

With new electric power plants planned to be operational in the near future, the need to deliver this new source of power will be required. The largest users of power will be those people living in large cities and their suburbs and possibly industrial sites producing hydrogen.

To accelerate the planning, design, fabrication, construction, and testing, the federal government needs to provide the leadership to complete a new or modified network for transmitting a greater amount of electrical power. Some of the current problems with building new transmission lines are listed below.[35]

- Congress has to re-regulate the electric power industry to place the emphasis back with utility companies producing growing volumes of reliable power and transmission capability. The Enron model of buying, moving, and marketing electricity across state lines must be eliminated. Deregulation of electricity markets, particularly along the eastern seaboard and in the southwest, has further burdened the electric grid as generators send their electricity hundred of miles to the highest bidder across wires originally intended for much more localized traffic.

- Our legislators want to change the mix of energy producers. The reason you need more lines is that the new power plants aren't going to be located where most of the existing plants are located now.

- Acquisition for the right of way to build new power lines is usually difficult in that it can be costly and may require a considerable expenditure of time and human resources. To obtain rights from unhappy individual property owners, local utility company and homeowner negotiation often leads to litigation. When utility power lines pass across state borders, a utility company is required to obtain approval from each state affected.

The challenges presented for the installation of new transmission line are similar to those discussed earlier for the installation of new gas pipelines.

Peter Huber of the Manhattan Institute suggests, "Building of a national grid, a financially modest undertaking for an industry as large as the power industry already is, will unleash innovation and competition on both the supply and demand side of our energy market. It should be built."[36]

What's Ahead?

There are a myriad of energy-related government-funded projects underway. Some projects have been funded with our tax money for over forty years. Some of the forty-year-old projects offer little more now, in terms of economical, practical energy production, than they did forty years ago. There appears to be no cohesive plan to integrate all of our energy-related initiatives. We need to focus on just a few of the many energy options being considered. Our government needs to select those few energy options that offer economic and effective solutions to meet the goal of energy independence.

Advocates for solar energy propose to cover hundreds of square miles with solar electric plants. They want to build hundreds of miles of new high-voltage direct-current transmission lines across a large portion of our country. Solar advocates propose methods for storing electric energy produced by new solar-electric plants. One method proposed involves using solar power to pressurize air and storing the air in underground caves or unused coal mines. Hundreds of miles of pipelines are proposed to carry the pressurized air to various newly modified or newly constructed air/gas-fired electric power plants scattered across our country. Solar electric power is an expensive choice for energy production compared to what we pay now for electricity. Adding to the cost of solar electricity are the costs required for storage and transportation of air to new air/gas-fired electric power plants.

Advocates for wind energy want to extend existing electric transmission lines into remote areas of our country where wind turbines are planned to be located. Wind advocates propose underground storage of pressurized air and hundreds of mile of pipelines as a means to store wind energy. Other wind advocates propose locating large arrays of wind turbines in the ocean or in our great lakes. Why do we continue to plan and advocate placement of wind turbines in the ocean and in our lakes? Offshore wind energy costs are shown to be more than double the energy cost of wind turbines located on land. Without government subsidies, the cost of land-based wind energy cannot compete or survive in our competitive energy market.

If we continue on the path chosen by our government, we will continue to financially subsidize the installation of more wind turbines, solar farms, and biofuel production plants. Our electric bills will increase because the price to produce energy from wind and solar is significantly higher than what we pay now for electricity. If our energy policies continue, the government will continue to spend ever-increasing amounts of our tax dollars to subsidize the increase in biofuel production. Biofuels are more expensive to produce than gasoline. The energy produced by wind turbines and solar conversion devices will not diminish our need for gasoline. Wind- and solar-produced energy will only slightly diminish the need for the natural gas and coal used by our electric power plants. If we allow it, our government will continue to spend our money to subsidize a shift from the low-cost energy sources of coal, oil, natural gas, and nuclear, to the higher-cost renewable energy sources.

An article in the Wall Street Journal, September 24, 2010, reported that General Electric Co. Chief Executive Jeff Immelt warned that the lack of a comprehensive US energy policy and its "stupid' current structure are causing America to fall behind in new energy fields. He indicated that we are experiencing a stalled effort to revamp US energy policy and that our country has failed to maintain and expand its nuclear industry.

If our energy policies continue during the next ten years:

- Our electricity costs will likely double.

- Our government will try to convince us that higher energy costs using renewable energy sources are worth it because it's "green power."

- We will use gasoline, ethanol, or natural gas in our cars for the next twenty years.

- We will see significant fluctuations in the price of gasoline and natural gas as we've experienced for the last forty years.

- We can expect very little change in our carbon dioxide emissions because the large majority of cars and trucks will still be burning fossil fuels.

- We will see very little change in the share of renewable energy production compared to our conventional energy sources. As a result, we can expect very small reductions in carbon dioxide emissions from our coal and natural-gas-fired power plants.

- We will continue to pay for government subsidies for biofuels, wind and solar energy, and energy conservation, to name just a few of the many subsidies.

- We will be forced to pay for grants and R&D contracts and continue to fund energy-related government agencies to search for economical and "clean" renewable sources of energy.

Because the large oil and gas companies, understandably, want to sell their products, they want us to continue to burn fossil fuels in our cars. We will continue to rely on fossil fuels because the oil and gas companies, with their influence, can convince or legislators that it's the right thing to do. Their position, quite correctly, is that renewable energy won't provide a significant amount of energy for at least twenty years. Our oil and gas industries conclude that fossil fuels must be used to bridge the gap from now until renewable energy is able to meet our energy needs. They advertise in favor of, and support continued research on and development of, renewable energy.

The electric companies know that renewable energy will not provide enough energy over the next ten to twenty years to alter our need for fossil fuels. We hear very little as to how much our electric bills will increase as renewable energy production increases. Will our elected representatives continue to pass legislation that favors their friends and large campaign contributors like the oil and gas companies? Or will our representatives change our energy policies so that our country can actually reduce our consumption of fossil fuels, reduce our electric bills, and clean up the air we breathe?

Our huge consumption of oil, gasoline, biofuels, and natural gas does not have to continue for the next twenty years. We don't have to continue to pay for research and technology development for the next ten to twenty years on a myriad of renewable energy alternatives. We have the right to direct our leaders to formulate a new energy plan that will better meet our energy needs. After all, they are supposed to be working for us! We can select energy sources that will supply all of the energy we need in the future. The energy sources will not cause air pollution and will eliminate our reliance on imported fossil fuels. Without the need to buy electricity, foreign oil, and natural gas, we can add billions of dollars to our economy rather than to spend it in other countries. The money saved can be used to invest here in our country. Selecting the most economical energy sources will reduce the price of our electric bills compared to what we are paying now.

What is this mysterious source of energy that our oil and gas companies never mention? More new electric power plants are needed if we seriously want to eliminate our dependence on foreign oil. With more nuclear and gas-fired electric power plants we can significantly reduce our carbon dioxide emissions. We have paid for most of the research and technology development needed to start building more new nuclear electric plants. Planning for about thirty new nuclear electric plants is underway in our country. However, we need at least ten times that many more new electric power plants over the next twenty years to eliminate our dependence on foreign oil.

Part Three

The Electric-Hydrogen-Powered Economy

Hydrogen Production

Hydrogen from Natural Gas

Hydrogen is the most common element in the universe and can be produced from readily available sources such as methane and water. Our hydrogen production methods are either inefficient or produce greenhouse gases. Hydrogen is produced today on an industrial scale by the petrochemical industry. Natural gas, with its content of hydrogen, is converted into hydrogen. Natural gas is used as the source for hydrogen, and additional natural gas is used to provide the heat necessary to perform the conversion process. Undesirable greenhouse gases are a byproduct of this conversion process.

The sustainable energy supply system of the future will use electricity and hydrogen as the dominant energy carriers. Hydrogen produced from off-peak nuclear-generated electricity will play an important early role in the transition to a hydrogen-based energy economy.[37] During off-peak hours, nuclear plants can generate more electricity than is needed to supply the grid, and because of this economy, the excess electricity can be used to produce hydrogen. Nuclear energy offers the potential to efficiently produce large quantities of hydrogen without producing greenhouse gases.

The Department of Energy is exploring a range of nuclear reactor designs that show promise for commercial-scale hydrogen production. An advanced nuclear system can supply high-temperature heat to a hydrogen-producing thermo-chemical or high-temperature electrolysis plant. The plant can operate at high efficiency and avoid the use of carbon fuels, such as coal and natural gas. These two high-temperature processes show the greatest promise for efficient economical hydrogen production. The hydrogen production system and heat-transfer components, such as heat exchangers, will require the development and test of high-temperature, corrosion-resistant materials. During 2009, the DOE selected a single nuclear-hydrogen-production technology for use with the next generation nuclear plant on which to focus development activities. The DOE will continue international collaborations through the Generation IV Forum Very-High-Temperature Reactor Hydrogen-Production Project Arrangement, chaired by our country.

Hydrogen from Water

Production of hydrogen from water can be divided into three categories:

- Electrolysis uses electric current to split water in two parts: hydrogen and oxygen.

- High-temperature steam electrolysis is a variation on conventional electrolysis. This method uses heat and electricity to produce hydrogen. The steam electrolysis process produces hydrogen more efficiently than conventional electrolysis of water. It uses less energy to make the conversion.

- Thermo-chemical water-splitting uses very high temperatures of about 1,800 degrees Fahrenheit to split water into its component parts. The DOE is developing an advance high-temperature nuclear reactor, referred to as a "demonstration plant," that will convert water into hydrogen and oxygen by 2015. The demonstration plant will be smaller in size than a commercial plant. However, the plant will simulate and establish the viability of a larger, commercial-sized plant.

The approach used now for producing hydrogen from nuclear energy employs off-peak nuclear-generated electricity and existing water-electrolysis-production technologies. More efficient techniques, such as thermo-chemical water-splitting cycles and high-temperature electrolysis using nuclear electricity and waste heat, can be achieved with temperatures in the range of 1,200 to 1,800 degrees Fahrenheit. These temperatures are too high for the currently operating nuclear reactors to achieve. The new reactors being developed by the DOE will attain sufficiently high enough temperatures to produce hydrogen from high-temperature steam electrolysis or thermo-chemical water-splitting.

Hydrogen produced by nuclear reactors through electrolysis or thermo-chemical water splitting processes is very pure compared to hydrogen produced from coal or natural gas. Fuel cells require very pure hydrogen, and the high purity of hydrogen produced from nuclear reactors meets this requirement.

Production of hydrogen from nuclear energy emits no greenhouse gas or other emissions. However, there are carbon emissions as a result of mining and transportation of uranium and the uranium-enrichment process. There are also emissions that result during the construction of the power plant and waste management. But remember similar undesirable emissions result for coal and gas power plants before they are operational. For example, emissions are the result of drilling, mining, transportation, and power plant construction. Even wind turbines and solar arrays have to be transported to their sites and require construction activities that will result in carbon emissions.

Uranium, the main fuel for nuclear reactors, is readily available from stable, friendly countries. In 2002, sixteen countries produced over 99 percent of the world's uranium. Major suppliers exist in our country, Canada, and Australia. Canadian and Australian uranium mines today make up over 50 percent of the world's uranium supply. Compared to natural gas, uranium is low in cost and less sensitive to price increases. One uranium pellet about the size of the tip of your little finger has the equivalent energy potential of 17,000 cubic feet of natural gas, 1,780 pounds of coal, or 149 gallons of oil.

The safety of our nuclear reactors has been a very high priority in their design and engineering. About one third of the cost of a typical reactor is attributed to safety systems and structures. The Chernobyl accident in 1986 was a stark reminder of the importance of safety. At Chernobyl in the Ukraine, thirty people were killed, mostly by high radiation levels, and thousands more were injured or adversely affected. The reactor at Chernobyl lacked the basic engineering provisions necessary for licensing in most parts of the world.

The long-term storage of nuclear waste is a major challenge facing a nuclear energy initiative. The Nuclear Waste Policy Act of 1982 specifies that radioactive waste will be disposed of underground in a deep geological national repository or stored in giant steel-and-concrete casks. Significant scientific research and development is still needed to satisfy long-term storage and safety requirements.

Sites for new nuclear plants in our country were designed to host four to six reactors, and most of these sites have never been fully utilized. Obtaining approvals for the development of new or existing facilities will present a significant challenge in the development of nuclear hydrogen production capabilities.

Brad Linscott

Nuclear energy is a viable primary energy source that offers the potential for producing hydrogen through a process that is economical and produces relatively minor emissions in the nuclear fuel cycle. Together, nuclear energy and hydrogen technology offer the potential to meet our energy security needs. A transition to a hydrogen economy using nuclear energy offers our most economical energy alternative. In the future, high temperature reactors will provide the necessary energies to produce large-scale quantities of hydrogen using high-efficiency, high-temperature electrolysis or thermo-chemical water-splitting cycles. The DOE seeks to develop high- and ultra-high-temperature thermo-chemical technology by 2015 to produce hydrogen. Overcoming negative public perception, long-term storage, safety, geographic location of sites, and further demonstrations of advanced reactors are needed. The successful completion of this effort will prove that producing hydrogen from nuclear energy processes is a viable option as the fuel for the future. Hydrogen costs will be competitive with gasoline costs at refueling stations or stationary power facilities.

Hydrogen Technology Development and Market Transformation

The DOE has developed a possible scenario calling for hydrogen technology readiness to be complete by 2015.[38] The goal is technology readiness of hydrogen production, delivery, storage, and fuel-cell technologies. The technology would enable automobile and energy companies to opt for commercial availability of fuel-cell vehicles and hydrogen fuel infrastructure by 2020. Our government may consider becoming an early technology adopter and could enact policies to nurture the development of an industry capable of delivering significant quantities of hydrogen to the marketplace. Industries role would become increasingly dominant as the market penetration of hydrogen increases into the energy market. Some of the milestones of this DOE scenario are:

- Initial market penetration of hydrogen can start in 2011 and extend into 2025. During the initial market penetration, infrastructure investment begins with government policies.

- Starting in 2015, expansion of markets and infrastructure for hydrogen can begin and extend out to 2035.

- In 2025, fully developed markets and infrastructure can begin and extend beyond 2040. For the fully developed markets and infrastructure, transportation systems, hydrogen storage, and distribution can be commercially available across the nation.

To support the DOE scenario for hydrogen technology readiness by 2015, a complimentary scenario is needed to build new electric power plants for the production of hydrogen. New power plants are needed to supply the emerging energy market with a sufficient supply of hydrogen. The electric power plant milestones are:

- Start nuclear, natural-gas-fired, and geothermal electric power plant designs and establish contracts for the construction of 150,000 megawatts of electrical power for a target on-line completion date of 2020.

- Establish site selection for each new electric plant and plan for upgrading existing or for new electric transmission systems.

- Start construction of additional nuclear, natural gas, and geothermal electric power plants in 2020, having a combined capacity of 150,000 megawatts with a target for on-line start-up by 2030.

- Assuming the DOE nuclear hydrogen demonstration facility successfully demonstrates commercial scale production of hydrogen by 2015, commercial-sized nuclear plants of this type can be initiated before 2020. These plants will provide part of the 150,000 megawatts of additional capacity needed by 2030.

The goal of the DOE's Nuclear Hydrogen Initiative is to demonstrate the economic, commercial-scale production of hydrogen using nuclear energy. This initiative is planned to lead to a large scale, emission-free, domestic hydrogen capability.[39] Hydrogen is currently used to support the production of chemical raw materials needed for industrial processes and liquid petroleum products. The DOE initiative will lead to a transition from fossil fuels to a hydrogen-production economy. Hydrogen will provide the backbone of our industrial and transportation sectors. From an industrial perspective, hydrogen is used in the refining of petroleum into transportation fuels and heavy crude oils that are deficient in their hydrogen content. Hydrogen is currently used to manufacture chemical raw materials for the plastic industry that makes thousands of consumer products. The direct use of hydrogen in the transportation sector of our economy will substantially reduce our need for foreign and domestic fossil fuels. A transition to hydrogen would greatly improve our national security, because it can eliminate our need for foreign oil. Assuming the DOE can successfully demonstrate that nuclear plants can successfully produce hydrogen, we can expect to produce it at a price well below what we pay now.

The infrastructure needed to supply hydrogen to fueling stations across the country is certainly a large concern. Transmission of hydrogen by pipeline is the most economical mode. Consideration should be given to the use of existing natural gas pipelines to carry hydrogen. As the need for natural gas diminishes in lieu of electricity, these pipelines may be usable to transport hydrogen at relatively low pressures. If the pipelines currently used to transport natural gas are not compatible for hydrogen, at least the land rights for new pipelines are in place.

Ground Transportation, Fuel Conservation and Highway Safety

In addition to increasing energy production, it is prudent to select an energy policy that will implement the conservation of our energy resources. Changes in our ground transportation policies can result in a significant decrease in our consumption of fossil fuels over the next ten years. Fuel conservation will help propel our country toward the goal of energy independence. For example, interstate highway speed limits should be reduced, and more efficient methods for speed limit enforcement are needed. Reduced vehicle speeds on our highways will result in a reduction of fossil fuel consumption and reduce pollution caused by fossil fuel combustion. An important side benefit will occur as a result of reducing and enforcing reduced speed limits. Automobile traffic statistics show that reduced speeds will result in a reduction of highway accidents and fatalities.

Increased use of freight and high speed passenger trains will reduce our consumption of fossil fuel. Fuel consumption for mass transit high speed passenger trains is considerably less than the fuel needed to transport an equal number of passengers in their automobiles. Similarly the fuel consumed by a freight train is less than the fuel consumed to transport an equal amount of freight by a number of tractor-trailers. An increase in the use of high speed passenger trains and freight trains will reduce automobile and truck traffic on our highways and reduce air pollution. Highway traffic statistics show that when the number, of automobiles and trucks traveling on our highways, is reduced the number of fatal and injurious accidents decrease.

Interstate Highway System

During the Eisenhower administration in July, 1954, Vice President Nixon announced the Interstate Highway program. The administration, the auto industry, the petroleum industry, the cement industry, the trucking industry, the manufacturers of earth-moving equipment, and the big civil engineering firms all agreed on the plans to build the new highways. The basic design for this gigantic undertaking was to build a 40,000-mile-long, coast-to-coast, border-to-border network of limited-access, grade-separated, high-speed divided highways that would connect all cities of more than 50,000 inhabitants.

Two years later, in 1956, Congress created the Highway Trust Fund to finance the construction of the Interstate Highway System.[40] This system, built in partnership with state and local governments, has become central to transportation in our country during the last fifty years. We set out in 1954 to build 40,000 miles of interstate highways. Now we have a National Highway System of over 160,000 miles of interstates and other roads of importance for homeland and national defense and mobility.

The Highway Trust Fund receives tax revenues allocated to the fund from a federal gasoline tax of 18.4 cents per gallon. If we consume 140 billion gallons of gasoline each year and we invest 18.4 cents for each gallon purchased, we pay about $26 billion each year into the Highway Trust Fund.

Like many other government programs, the federal role in surface transportation has expanded to include broader goals and more programs. In 2005 the Safe, Accountable, Flexible, Efficient Transportation Equity Act authorized $243 billion over five years for highways, highway safety, and public transportation. The $243 billion authorization includes monies from the Highway Trust Fund and from the General Fund of the United States Treasury even though the Treasury is financially operating in the "red."

The majority of the $243 billion authorized for fiscal years 2004 to 2008 was spent to construct and maintain highways and bridges.[41] However, the act specifies that a portion of the $243 billion can be used for purposes other than the construction and maintenance of highways and bridges. The funds are being used for safety, metropolitan planning, transit, and transportation enhancement activities such as trails for transportation purposes, pedestrian walkways, bicycle lanes, parking, and other non-highway-related projects.

The Department of Transportation oversees four administrations: the Federal Highway Administration, the Federal Transit Administration, the National Highway Safety Administration, and the Federal Motor Carrier Safety Administration. The four administrations are responsible for administering the programs funded by the Highway Trust Fund and the United States Treasury. During the five-year-period, about $78 billion was obligated for purposes other than construction and maintenance of highways and bridges. An obligation by the government is a legal commitment that creates a legal liability of the government for payment.

- The Federal Highway Administration obligated about $28 billion for projects including: $2 billion for "Facilities for Pedestrians and Bicycles", $8 billion for "Safety", $4 billion for "Other", $3 billion for "Planning" and $1.8 billion for "Traffic Management Engineering".

- The Federal Transit Administration obligated, over the five-year period, $44 billion, mainly for formula and bus grants and capital investments grants/discretionary grants.

- The National Highway Traffic Safety Administration obligated $3 billion that included grants for state and community highway safety programs, alcohol-impaired driving countermeasures incentive, and safety belt performance. They also obligated close to $1 billion for behavioral and vehicle safety research.

- The Federal Motor Carrier Safety Administration obligated about $2.4 billion during the five-year period. They obligated almost $1 billion for a Motor Carrier Safety Assistance Program grant and $0.75 billion for operating expenses.

For various reasons, the Highway Trust Fund is headed for bankruptcy and, in 2008, the Department of Transportation told states the shortfall will lead to a slowdown in transportation funding.

From 2001 through 2009, Highway Trust Fund expenditures exceeded collections. Before 2001, the Highway Trust Fund was financially stable. The general fund of the Treasury, for the most part, has run large deficits each year. The transfer of money from the general fund to highways increases the general fund deficit. The Highway Trust Fund was established with a clear objective—to build new highways for high-speed ground transportation and to maintain them.

Fuel Conservation and Safety on Our Highways

When you drive your car, you impart the single largest impact on earth affecting a climate change. For each gallon of gasoline a car engine burns, twenty pounds of carbon dioxide is produced and added to our atmosphere. In addition to carbon dioxide, quantities of carbon monoxide, oxides of nitrogen, and sulfur are added to the air we all breathe. Our cars and trucks burn 380 million gallons of gasoline each day. Burning 380 million gallons of gasoline produces about 7 billion pounds of carbon dioxide that is disbursed into our atmosphere in one day. For the year 2007, the combustion of gasoline produced nearly 1,300 million tons of carbon dioxide.

Pollution is easily reduced by reducing the amount of fuel consumed by combustion. The amount of fossil fuel consumed by motor vehicles can be reduced by reducing and enforcing speed limits on our highways. A significant reduction in highway accidents occur as speed limits are reduced and enforced.

Motor vehicle collisions are one of the leading preventable causes of death in our country. Driving your car is one of the deadliest undertakings in the history of mankind. In 2005, about 33,000 people were killed in motor vehicle traffic crashes and 2.4 million people were injured due to automobile related accidents.[42] The monetary result of these fatal and nonfatal unintentional injuries affects every American household, causing higher prices for goods and services and higher taxes. It is notable that the penalties for killing and injuring with motor vehicles are often very much less than for other actions with similar outcomes. The number of fatal auto-related injuries declined from 33,000 people in 2005 to about 27,000 people in 2008.

Why do our newspapers provide daily accounts of casualties from the Iraq and Afghanistan war? It certainly reminds us that a terrible war is going on, and we would like to have the war over and our soldiers back home. On July 7, 2009, 4,322 members of the US Military and Defense Department civilians were killed over a period of six years since the start of the Iraq war in March, 2003, as shown in Figure 1.

Tuesday July 7, 2009

7 troops killed in Afghanistan

U.S. military deaths

Iraq
Updated: Monday

4,322
Members of the U.S. military and Defense Department civilians killed since March 2003.

3,460
Military personnel killed as a result of hostile action.
There were no new deaths or identification reported by the military.

Afghanistan
Updated Monday

642
Members of the U.S. military killed since late 2001.

475
Military personnel killed by hostile action.

68
Members of the U.S. military killed outside the Afghan region in support of Operation Enduring Freedom.

Washington
Highway deaths at a record low
Deaths on U.S. highways have dropped to a record low during the first six months of 2009, continuing a recent trend of fewer people dying on roads. the National Highway Traffic Safety Administration reported Friday that 16,626 people died in traffic crashes between January and the end of June, a 7 percent decline from the same period last year. It followed up on a record low number of deaths achieved for that period in 2008, when an estimated 37,261 motorists died, the fewest since 1961.

Figure 1: Comparison of Typical Newspaper Report on Military War Casualties and Auto Related Civilian Fatalities

Brad Linscott

What if our newspapers started to report to us, each day, the number of people killed in automobile- and truck-related accidents? The National Highway Traffic Safety Administration reported in October, 2009, that 16,626 people died in just six months due to traffic crashes between January 2009 and the end of June 2009. (See Figure 1.) About 5,700 people are killed in auto accidents every nine weeks compared to the human death toll of 4,322 during six years of war in Iraq and Afghanistan.

Many people, including our representatives in Congress, have voiced their concerns about our soldiers that are being killed in Iraq and Afghanistan. We all want to stop the killing and maiming that wars claims. But why aren't these people and our congressional representatives voicing even more concern over the slaughter on our highways? Over 16,000 people were killed while traveling our highways in just the first six months of 2009!

A large number of auto accidents involve tractor-trailer rigs and automobiles. Even with the latest safety technology used in auto manufacturing, our cars are no match for a large truck when an accidental collision of the two occurs. Large trucks traveling the Interstates can be perceived as railroad freight cars on rubber tires. The large tractor-trailer trucks, often exceeding the posted speed limit, possess a huge momentum compared to the standard passenger cars. The impact between a large truck and the standard automobile can result in severely damaging results that cause death and injury.

It appears that our state and federal government walk a narrow path between enacting measures to reduce auto-related death and injury while avoiding measures viewed by drivers as too intrusive. Our federal and state legislators need to take action to reduce auto-related fatalities. The trucking industry provides campaign money and supports lobbyists to influence our legislators to enact laws that favor their industry and at the same time can jeopardize the safety of people using automobiles on our highways. For example, interstate highway speed limits have been allowed to increase in some states because the trucking industry has convinced our legislators that higher trucking speeds are needed. Department of Transportation statistics show that the reduced speeds during the 1970's government-imposed speed limits were responsible for a reduction in highway fatalities.

A significant safety measure would be to designate automobile only roads and interstate highway lanes. The New Jersey Turnpike has "car-only" lanes and mixed car-and-truck, traffic lanes. Accident rates are found to be higher—in some cases as much as four times higher—on mixed traffic highways compared to car-only highways.

Have we decided to forget about the fact that our highways provide a "killing field" for motorists? Or is there something we can do to reduce the number of people being accidentally killed on our highways and interstates? One effective method is to reduce the number of vehicles traveling on our roads and interstate highways. During 2008, road fatalities fell rapidly as Americans cut back sharply on driving because of record high gas prices.[43] Here are some things we can do to reduce fossil fuel consumption and at the same time reduce the number of auto-related deaths and injuries.

- Compared to automobile transportation, high speed passenger trains use less fossil fuels and produce less pollution per passenger mile. As high speed passenger train rider ship increases, the number of automobiles on our highways will decrease. A decrease in the number of automobiles on our highways will result in a decrease in the number of auto related accidents.

- Compared to trucks, rail transportation of freight requires less fossil fuels and produces less pollution. We can modernize our freight-train systems to provide economical and rapid transportation for our trucking industry that would reduce the number of trucks on our interstate highways. Highway fatalities will decline as the number of trucks traveling our highways are reduced.

- Enforcement of highway speed limits will result in a reduction in the consumption of fossil fuels and produce less pollution. France has successfully reduced their automobile accidents by introducing an automated speed-enforcement system. Our country should begin to install speed-control systems similar to those used in France.

High-Speed Passenger Trains Will Reduce Fossil Fuel Consumption

The passenger-train business in our country has seen very little passenger-train activity in the last thirty years.[44] As a result, there is no passenger-train industry and hence no industry experts. To bring the passenger-train service into the twenty-first century will require federal funding, planning, and implementation.

Our tax money has been used to invest in air transportation, the expansion of ocean ports, and in building interstate highways while passenger-rail transportation has been essentially neglected. Billions of federal dollars have been spent to support our highways, airways, and waterway systems. Why not high-speed passenger-train service as well? The need for high-speed passenger-train service between large cities in our country is long overdue.

We, the taxpayers, have invested hundreds of billion dollars for infrastructure needed by the airline system. In 1992, it was estimated the replacement value of our commercial airline system was valued at $1.0 trillion. The majority of this investment was paid for with federal grants and tax-free municipal bonds. In 1988, the Federal Budget Office found that in spite of user fees paid into the airport and Airway Trust Fund income, the taxpayer still had to pay $3 billion in subsidies each year to support the Federal Aviation Administration's (FAA) expenditures. In 1988, the expenditures were needed to maintain the FAA's network of more than 400 control towers, 22 air traffic control centers, 100 radar-navigation aids, 250 long-range and terminal radar systems, and its staff of 55,000 traffic controllers, technicians, and bureaucrats.

Cruise ship travel started in the 1960s as a mode of travel mostly for people in the upper income levels. The federal government has spent billions of dollars, allocated to local communities, to improve their deep-water ports. As a result, the cost of cruise-ship travel began to decline and the number of cruise-ship passengers increased as did the number of new cruise lines. What was formerly considered a luxury for the super-rich has now turned into a virtual entitlement for middle-class America.

Compared to Europe and parts of Asia, we are many years behind in the development of high-speed rail service.[45] The Eurostar Rail connects London, Paris, and Brussels at speeds up to 186 miles per hour. Japan's bullet train travels at 180 miles per hour and they are investing heavily in a train that will exceed 200 miles per hour. Taiwan has trains capable of reaching 186 mile per hour. The only high-speed passenger train we have in service is the Acela. It operates in the Boston/New York/Washington Northeast Corridor. The Acela can travel up to 150 miles per hour, but averages less than 86 miles per hour between Washington D.C. and New York City. The train line speeds are restricted because of poor infrastructure and track conditions.

Passenger trains require federal infrastructure investment in a modern right-of-way and a modern command-and-control technology just as cars and airplanes do. But today, after nearly a century of federal investment in road, air, and water infrastructure, we still have no federal funding for the development of a modern passenger rail system. Passenger trains capable of speeds between 150 and 200 miles per hour are needed. High-speed train routes must be selected that will provide direct competition with "short hop" air travel and interstate automobile transportation.

The driving distance between Los Angeles, CA and San Francisco is about 380 miles. For air travelers this is a short hop. High speed passenger train service between these two cities is ideal because it will provide travelers with an alternative to using the automobile or the airplane. California received $2.25 billion in stimulus money to help connect Los Angeles and San Francisco with passenger trains that will travel at speeds up to 220 miles per hour.[46] An estimate of more than $40 billion is needed to complete the high-speed passenger train link within ten years.

The flight time from Cleveland, Ohio, to Chicago, Illinois, is about one hour. Boarding an airplane requires two hours, and deplaning requires about one half hour. The total travel time by airplane is three and a half hours. A high-speed passenger train traveling at an average speed of 150 miles per hour can cover the 300-mile distance in two hours. Allowing a half hour for boarding and a half hour for departing, the train would save the traveler one half hour when compared to the total travel time to fly from Cleveland to Chicago. During inclement weather conditions at our airports, large numbers of passengers are required to wait for safer flying conditions. The high-speed passenger trains are usually not restricted by local weather conditions. The availability of a modern train service can provide the airline traveler grounded by weather conditions a choice. The traveler can board a high-speed passenger train for transportation to their destination, or spend the day or night at the airport terminal waiting for the weather to change.

There are many people whose lives can be improved with the availability of high-speed passenger service. For example, it would assist college students traveling between home and school and people visiting their families. High-speed train service would be beneficial for people relocating to new jobs or checking out an out-of-town job opportunity and professional groups heading to a conference. High-speed train transportation would be ideal for foreign visitors who wanted to see our country close up and meet Americans on route and would benefit retirees that wanted a relaxing and informative travel experience.

Modern train terminals are needed at each city selected. Each new train terminal should be modeled after the airline terminals. It would be nice to be able to rent a car, turn in a rental car, or have adequate parking available for personal cars at a train station. How about some fueling stations for cars powered by either electricity or hydrogen? Of course, like the airline terminals, it would be nice for train stations to have easy access to public transportation such as buses and commuter trains, restaurant service, modern and clean public restrooms, and easy access to hotel/motel accommodations. These passenger train accouterments are needed to provide the traveling public with an attractive alternative to air and automobile transportation for travel distances of 500 miles or less. New terminals will attract private investment and much-needed employment opportunities.

The largest challenge to bring passenger trains into the twenty-first century is the need for new, or the need to modify existing, roadbeds. Modification to interstate highways to accommodate high-speed passenger trains is an option. The interstate highway system may provide an attractive alternative for the design and construction of new roadbeds for high-speed trains compared to the modification of existing railroad roadbeds. The need to expand automobile highways and interstates can be eliminated with the introduction of high-speed passenger train and modernized freight-train service.

Interstate Speed Control will Reduce Fuel Consumption, Pollution and Accidents

France introduced an automated speed-enforcement system for their highways in 2003, when France's President Jacques Chirac declared road safety a major national priority for their country. The new initiative broke away from the past practice characterized by random ineffective speed limit enforcement. Beginning in 2003, 500 new radar speed traps have been installed each year on their highways. By the end of 2007 1,850 devices were fully operational, and by 2008, over 2,200 were operational.

The French system uses fixed speed measurement units that use bifrequency Doppler technology with digital cameras to obtain quality images of automobiles and trucks traveling their highways.

Photographs of offending vehicles are taken of the front and rear portions of the vehicles. In 2008, the system notified 17 million offending vehicle owners. The new operating system is credited with reducing the number of auto-related fatalities by 40 percent in 2008 compared to 2002. French authorities attribute the new system with saving 12,000 lives since 2003, when the first part of the system was installed. The number of auto-related injuries have diminished, although no information is available for the purpose of comparison.

The automated system uses fixed units and mobile units. The mobile units are used by police, at their discretion as to time and place. French authorities have chosen to provide signs to alert motorists before the location of a fixed detection unit. Mobile detection uses unmarked, specially equipped cars that are used less and randomly, compared to the fixed units. The mobile units require the presence of police officers. The mixed strategy achieves both general and specific deterrence. The mobile units provides for numerous checkpoints on roadways and generates a situation of widespread uncertainty for offending drivers.

In 2010, the country of Belarus installed 110 automated speed-control units on their expressway. The units were installed to improve road safety along the expressway between Poland and Russia. The deployment of the custom-designed system will enable the Belarusian authorities to rapidly process all speeding violations, as drivers will have to pay their tickets when they attempt to leave the country. It is expected that the new system will notably improve safety on the M1 expressway, which the Belarus government authorities consider to be an accident black-spot.

City Bypass Beltways

Beltway Traffic Control will Reduce Fuel Consumption, Pollution and Accidents

While traveling on our interstates beltways that surround large cities, have you seen a tandem line of cars traveling bumper to bumper at or above the speed limit? This scene is common on interstate beltways that bypass the downtown area of large cities. The tandem line of cars may remind one of the NASCAR cars as they race around the oval track. One major difference is that the NASCAR drivers are trained and experienced in the art of high-speed, bumper-to-bumper driving. Most drivers traveling on the beltways have not had formal training for this kind of driving. Another difference is that the NASCAR drivers have more sophisticated safety equipment than our standard cars. As a result, they are better protected from serious injury in the event of a car crash.

The technology is available and, if implemented, can be used to identify the people involved in this type of driving misbehavior. Those offenders can be notified and presented with an appropriate traffic fine. All of the offenders can be identified and sited without having a police car give chase.

Chuck Hurley, chief executive of Mothers Against Drunk Driving, recalls that half of the decline in road deaths during the 1970s was attributed to high gas prices. The remainder was linked to the lowering of freeway speed limits to fifty-five miles per hour.

City Traffic Control will Reduce Fuel Consumption, Pollution and Accidents

The Northeast Ohio Area-wide Coordinating Agency recently reported their results of a three-year study on accidents that occurred at high-volume intersections in north central Ohio.[47] The study, started in 2005, was concluded at the end of 2007. A goal of the report is to help planners to review car accidents at a large number (over 100) of intersections and to take action that will improve the safety at these intersections. At a heavily traveled intersection located in downtown Cleveland, Ohio, the report provides this information for the three-year period: One hundred and twenty-six automobiles collided, causing one human fatality and forty-three human injuries. The average number of vehicles traveling through the intersection was 53,822 each day. The average cost to repair each vehicle involved in the 126 collisions was about $390. The total repair bills amount to about $49,000 over the three years just for this one intersection.

Federal and state government help is needed to stop lengthy court litigation over the legality of video systems to ticket speeders. For example, the city of Cleveland faces a possible avalanche of court challenges to tickets issued by the city's controversial traffic cameras after a lawyer won an unprecedented court ruling in 2009.[48]

The use of a centralized, video-monitored, traffic-flow and speed-control computer system is needed to smooth the flow of traffic on heavily traveled urban and city streets and at high-flow intersections. Such a system could significantly reduce the number of fender benders, human injuries, and fatalities. On-demand entry from secondary streets and business establishments onto heavily traveled streets can be eliminated during heavy traffic flow. Computer-controlled traffic flow and timed traffic lights on the main streets can be maintained during heavy daytime traffic flow.

With new traffic controls in place, a significant reduction in automobile and truck fuel consumption can be attained. Energy savings can also be realized by reducing the number of accident related repairs and the need for replacement parts.

People's lives are at stake while driving in large cities. The technology for traffic control has not significantly changed since the early 1950's. Our local, state, and federal governments should support a stronger effort to use modern traffic control technologies. Implementation of modern traffic control technologies can provide the public with safer transportation by automobile and at the same time reduce fuel consumption and pollution.

Hydrogen Fueled and Electric Automobiles

We have the opportunity to eliminate the need for imported fossil fuels by starting to use, and to promote using, electric/ battery powered automobiles for city driving. For long distance highway transportation the use of hydrogen as a fuel can easily replace the need for fossil fuels. Combustion of hydrogen produces zero pollution. The consumer cost of hydrogen compared to gasoline is expected to be at least 50 percent less.

For automotive transportation, three primary types of vehicles should be designed to meet consumer needs. Each vehicle is capable of pollution-free operation without the need for fossil fuels. The three types proposed are:

- Electric/Battery-powered vehicles: These vehicles are ideal for urban transportation, where relatively low mileage trips are required.
- Hydrogen-fueled vehicles: These vehicles can be used for long-distance transportation, with hydrogen fuel stations available for rapid refueling.
- Hybrid Electric/Battery/Hydrogen-Fueled Vehicles: The hybrid vehicles can be used for both urban transportation and long-range travel.

With adequate amounts of hydrogen as a fuel, the need for small, lightweight, and more injury-prone passenger vehicles can be eliminated.

Electric/Battery-Powered Automobiles

Appropriate government departments, including the Department of Energy and the Department of Transportation, must continue to work with our automakers to convert the production of gasoline-powered automobiles to the production of electric/battery-powered vehicles. Electric/battery vehicles are limited to a short range of travel. Typically today's electric car can travel about 100 miles before battery recharging is required. The increase in electric/battery automobile production must be coordinated with the increase in electric power capacity and the transmission of electric power to automobile owners. It is envisioned that short-range electric/battery vehicles are best suited for city and suburban transportation needs. Most city/suburban workers are commuting about 100 miles per day. An overnight recharge for the short-range vehicles is needed.

Andrew Grove, former chief executive of Intel Corp., proposes our aim to be that one trillion vehicle miles per year (about a third of the total) be fueled by electricity within a decade. [49] President Obama stated his goal of 1 million electric vehicles on the road by 2015 during his 2011 State of the Union address.

Hydrogen-Fueled and Hybrid Electric/Hydrogen-Fueled Vehicles

Hundreds of hydrogen cars are on the road around the world.[50] Vehicle manufacturers, including Honda and BMW, are putting hydrogen vehicles into the hands of consumers for extended test drives. The cars have hydrogen engines or run on electricity from fuel cells. The DOE, car manufacturers, and the National Hydrogen Association say consumers can expect to see hydrogen vehicles in auto showrooms by 2020.

The transformation to a hydrogen economy will serve at least two major objectives in the international area.[51] First, the reduction in oil imports with the attendant increase in energy independence is a clear goal for our country, and hydrogen will be a strong contributor. Second, if our country can forge a lead in hydrogen technologies, a global competitiveness will be fostered. The movement to hydrogen in particular could well be an opportunity for our auto firms to recapture market share lost to foreign multinationals during the last ten years.

Hydrogen internal combustion engine efficiencies of 45 percent have been projected based on experimental engine data. Internal gasoline combustion engines with electric spark ignition have thermal efficiencies of less than 32 percent. Most of the experimental and theoretical effort supports fossil fuel combustion and emission control.[52] Of the fifty projects reported, only two projects (4 percent of the total effort) are investigating the use of hydrogen as a fuel. If the decision is made to move forward into the hydrogen economy, much of the fossil fuel combustion effort can be scaled down and eventually eliminated.

To lubricate all of the moving parts in our automobile engines, motor oil is added into the engine crankcase. To reduce gasoline combustion gas emissions into the atmosphere, a portion of the emissions are vented into the crankcase of our engines. The engine oil has to be replaced periodically, not because the oil wears out, but because the exhaust pollutants added to the oil can reduce the operating life of the engine parts. By burning hydrogen in our automobile engines, the same engine oil can be used for the lifetime of the engine because water vapor, the only combustion emission, is vented into the atmosphere.

Appropriate government departments must work with our automakers to help foster the conversion from fossil fuels to hydrogen for trucks, buses, and long-range automobiles. The production goals of hydrogen-fueled vehicles must be coordinated with the increased availability of electrical power and of hydrogen. Fuel stations must be established to provide gaseous hydrogen to consumers.

Fuel Cells for Automobile Ground Transportation

For transportation applications, fuel-cell systems could substantially reduce the nation's dependence on imported petroleum and emissions of carbon dioxide and other pollutants resulting from the combustion of fossil fuels. Fuel-cells systems produce only water and heat as by products. The fuel cells themselves produce no emissions of carbon dioxide or other pollutants. For the case where hydrogen is produced from natural gas, the resulting emissions are expected to be 50 percent less than emissions from advanced gasoline hybrid vehicles by 2020.

Applications for fuel cells that are currently commercially viable or are expected to achieve viability in the near term include vehicles used for material handling and airport ground support. Other applications include auxiliary power units, primary power systems, and portable power supplies. The Department of Energy's FY 2010 Congressional Budget Request requested $63 million to continue fuel-cell systems research and development.

Although fuel cells used to power automobiles stand to provide the greatest benefits, they also face some of the steepest challenges that include:

- A decrease in the cost for fuel cells is needed and their durability under varying operating conditions must be improved.

- A significant investment in infrastructure is needed in order to support widespread use of fuel cells used to power automobiles.

- A well-refined manufacturing capability is needed in order to compete with the facilities built to mass produce the gasoline-fueled internal combustion engine.

A Department of Energy grant funded a report that casts some doubt as to the feasibility of using the fuel cell for light-duty-vehicle applications in the near future. The federal government has been active in fuel-cell research for over forty years. In spite of substantial R&D spending by the DOE and industry, fuel-cell costs are still a factor of ten to twenty times too expensive compared to the more standard available options for producing electric power. The fuel cells are short of required durability and their energy efficiency is still too low for light-duty-vehicle applications. Accordingly the challenges of developing the *proton exchange membrane*, fuel cells for automotive applications are large, and the solutions to overcoming these challenges are uncertain.

Because of uncertainties surrounding the success of the fuel cell, emphasis needs to be placed on the development of hydrogen/electric hybrid light-duty-vehicles. The vehicle should use the reliable and time-proven internal combustion engine designed to burn hydrogen. The Toyota Prius Hybrid automobile, now in production and on the road, can travel on the average of forty miles on one gallon of gasoline. The Prius was selected, as an example, to show that it can easily be modified to travel 360 miles when equipped to burn seventeen pounds of hydrogen gas.

Scenario for Hydrogen Fueled/Electric Hybrid Automobiles

One gallon of gasoline weighs about 5.6 pounds. This amount of hydrogen gas at ground-level atmospheric pressure has a volume of about 1,000 cubic feet. If this volume were enclosed within a square cube, each edge of the cube would be ten feet long. By applying external pressure to the gas, the 1,000 cubic feet of hydrogen can be compressed to a smaller volume. For example, if 1,000 cubic feet of hydrogen is compressed by applying a pressure of 1,000 pounds per square inch, the volume will be reduced to 14.7 cubic feet. The 5.6 pounds of hydrogen can be stored in a fuel tank having an internal volume of 14.7 cubic feet when pressurized to 1,000 pounds per square inch.

If the shape of the fuel tank is cylindrical with an internal diameter of twenty-four inches, a cylinder length of fifty-six inches is needed to contain 14.7 cubic feet of hydrogen. Assuming the fuel tank is constructed of aluminum alloy, a wall thickness of about 0.6 of an inch is needed to safely withstand the inside pressure. The aluminum fuel tank would weigh about 300 pounds. If carbon fiber composite materials were used to construct the tank, the tank would weigh less than 150 pounds.

The energy contained in hydrogen by weight is almost three times (120/44) that of gasoline. If the Prius was fueled with 5.6 pounds of hydrogen, it could travel 120 miles. If three of the hydrogen fuel tanks were installed in the Prius, it could travel about 360 miles. A travel distance of over 300 miles is comparable to most gasoline-model automobiles on the road today.

Vehicle Design Integration with Fuel Tanks

One aluminum fuel tank, sized at twenty-four inches in diameter, fifty-six inches long, and weighing 300 pounds, requires a considerable amount of vehicle volume and is heavy. Three fuel tanks will more than fill up the trunk compartment of the Prius. In addition, 900 pounds located in the trunk compartment of the Prius could adversely affect the center of gravity and balance of the vehicle.

Hydrogen fuel tanks can be manufactured as part of the automobile under-frame structure. With a small amount of additional material, the tanks can easily carry the frame structural loads. The advantage of this concept is that the car doesn't have to "carry" the hydrogen tanks separately. They won't take up trunk space. Reciprocating internal combustion engines are proven to be reliable and can easily be modified, by design, to burn hydrogen instead of gasoline. The hydrogen-fueled automobile must be structurally designed to accommodate and safely operate with fuel tanks loaded with pressurized hydrogen gas.

Pressure vessels designed to contain hydrogen at pressures that range from 5000 to 10,000 pounds per square inch are not needed as others have proposed. Hydrogen fuel tanks pressurized from 5,000 to 10,000 pounds per square inch are more hazardous than tanks pressurized at 1,000 pounds per square inch when carried by a passenger vehicle.

Our country has the opportunity to become a world leader in the economical production of hydrogen. In addition we can provide leadership by demonstrating an economical method of providing energy that is nonpolluting. Our auto makers have an opportunity to manufacture automobiles and trucks fueled with hydrogen and powered by battery supplied electricity. These innovations can lead the world away from fossil fuels and provide clean energy. The United States has an opportunity to be the world's supplier of economically produced hydrogen and significantly reduce the world's need for crude oil.

Brad Linscott

Research

During the 1970s, future energy planners predicted that by the year 2000, the nuclear fusion process would be developed to the point where economical generation of energy would be demonstrated. Progress to develop a thermonuclear reactor capable of producing more thermal power than it takes to operate the reactor has taken much more time than anticipated.

The idea for nations to work together to harness fusion started in 1985, during discussions between President Ronald Reagan and Soviet leader Mikhail Gorbachev.[53] As a result of the discussions, our country, the Soviet Union, the European Union, and Japan began designing a fusion reactor. By 1998, there was general agreement, but several factors caused our participation in the project to unravel. Our country withdrew from the International Thermonuclear Experimental Reactor in 1998. In 2003, our country rejoined the partnership with three new members: China, South Korea and India. The pact called for the parties to share the reactor's $13.4 billion total cost of $7 billion to build it and the remainder to operate the reactor for twenty-five years. Design upgrades, proposed in 2008, boosted the total cost estimate to as much as $15.5 billion. The arrangement shares knowledge so that if sustained fusion works, each participating member can build its own reactor. Also if one member quits, it won't cripple the project. In 2007, with Congress and President George W. Bush were at odds over spending for 2008. Lawmakers unexpectedly, as a result, cut funding for many programs, including slashing the $160 million that we had pledged for the project to $10.7 million.

The mainstream approach and the one to be used for the International Thermonuclear Experimental Reactor is to suspend the plasma in a magnetic, donut-shaped tube called a *tokamak*.[54] The first tokamak was built in the Soviet Union in the 1960s. There are now several in our country and others are located around the world. As new tokamaks have been built, they have grown ever bigger and come ever closer to showing that the design could ultimately generate more power than it requires to operate. The largest one operating today is the Joint European Torus, or JET, located in the United Kingdom. It can produce up to thirty megawatts of thermal power—the first fusion device to put out as much thermal energy as it takes in. But its burn falls far short of the sustained one needed for true fusion energy. That calls for a larger tokamak with larger plasma, one that surely harbors surprises for scientists. The Experimental Reactor's plasma, at 28,250 cubic feet, will be ten times the size of JET's, and it will produce 500 megawatts of thermal power. The goal is to obtain ten times as much energy out as goes in. Construction is scheduled to be completed in 2016, and it will run experiments for twenty years. If the experimental reactor is successful, the next step will be to build a fusion reactor integrated with a power plant to make electricity.

Energy Independence—Expedite or Impede

Federally funded energy projects should be directed toward technologies that clearly demonstrate a capability to significantly reduce our dependence on fossil fuels. It's time to change the direction of our federal energy program. We need to select well-proven technologies that, when implemented, will significantly reduce our dependence on fossil fuels and provide economical electric energy to customers. Our country needs a new and focused energy program that will have lasting and positive effects on the economy in the years ahead. The right kind of energy program will create thousands of new jobs in all areas of private industry. The new job areas include energy-directed research, development, engineering, manufacturing, construction, testing, new computer systems, and operation of new kinds of equipment. Support for only a few selected energy research projects will lead to improvements in the technology to provide alternate methods for producing clean, economical energy in the future.

With the implementation of well-proven energy technologies, we can expedite attaining our goal of energy independence and provide economical electric energy to customers. The technologies that impede our goal exhibit little hope to reduce our dependence on imported fuels by 2020. Continued government support of energy technologies that are impediments will increase our cost for energy above the 2009 energy cost levels.

Electric Power Plants Will Expedite Energy Independence

The price of electricity increased from 7.6 cents per kilowatt-hour in 2004 to 8.9 cents per kilowatt-hour in 2006. The average retail price of electricity increased from 8.9 to 9.1 cents per kilowatt-hour in 2007.

In 2007 fossil fuel prices increased 7.0 percent, while the average retail price of electricity increased by only 2.6 percent. Our electric industry should be commended for maintaining relatively modest increases in our electric energy prices compared to the increased price for fossil fuels.

Total electricity sales include residential, commercial, industrial and transportation. In 2006 total electricity sales increased by only 0.2 percent from 2005. When all electricity sales are considered, sales increased by 2.8 percent during 2007. In response to the 2.8 percent increase in sales during 2007, our power plants were able to provide only 2.3 percent of the 2.8 percent of the increase in energy needed to meet the demand. Our electric power plants achieved the 2.3 percent increase by increasing the performance of existing coal fired, natural gas fired and nuclear capacity. Each of the three types of power plants set new production levels and increased average capacity factors in 2007, and yet we were unable to meet the total power needed.

But how did our electric utility industry make their supply meet the demand? They had to import electric energy from Mexico and

Canada. Our generating capacity is not keeping up with our daily demand for electric energy. Wind and solar energy cannot be relied on to meet baseline and peak electric demands until a stored energy infrastructure is established. A stored energy infrastructure for wind and solar energy will cost billions of dollars, and will require at least ten years before it can be operational.

Electricity is the key to our energy independence, and a national market for electricity is the answer to the global market for oil. If our country is serious about eliminating our dependence on fossil fuels, new electric power plants have to be built and put into operation during the next twenty years. The technology to support this effort is mature and efforts to support this goal can begin today.

How Many Electric Power Plants Do We Need?

We consume about 140 billion gallons of gasoline each year. Each day we consume 380 million gallons or about 15.8 million gallons of gasoline each hour. One gallon of gasoline has the electrical energy equivalent of 33.4 kilowatt hours. The electrical energy equivalent of 15.8 million gallons of gasoline per hour is about 530,000 MWh per hour. How many power plants do we need to supplant 140 billion gallons of gasoline used in our country over a one-year period?

Power plants are typically rated in terms of their electrical power. For example, a plant having a power rating of 1,500 megawatts is capable of producing 1,500 megawatts per hour. As a result, about 350 electric power plants, with a combined average rating of 1,500 megawatts each, are needed to supplant the energy we consume from gasoline each year. A power plant size of 1,500 megawatts was selected as the average size for a large power plant. In reality, the plant size may vary depending on its locality and other factors that will determine the capacity needed.

More detailed analyses are necessary to determine the exact number of power plants needed. Remember, 350 new power plants would only replace the energy consumed from gasoline used in our country. Additional power plants, over and above those currently operating, are needed to meet the demands of new industries, growth in population, and to replace aging plants that will be shut down.

General growth estimates for additional energy needs each year are somewhere between 1 percent and 2 percent per year. In forty years, if our growth rate is 1 percent each year, our energy needs would be 50 percent higher than our current capacity. If our growth rate, in terms of electrical capacity, is 2 percent each year over the forty-year period, our energy needs would be 2.2 times higher than our current capacity.

Our electrical energy needs will require about 350 new electric plants to replace the fossil fuels we use for ground transportation. And we will need an additional 350 new plants within the next forty years just to maintain sufficient energy production to meet a 1 percent growth in energy.

Nuclear Electric Power Plants Will Expedite Energy Independence

Do you recall the British Petroleum (BP) advertisement from 2009? They propose six alternative energy solutions. Their alternatives are oil, natural gas, wind, solar, biofuels, and the suggestion to increase efficiencies in our use of energy. Why is BP advertising these energy alternatives as the solution to meet our future energy needs? Is it because they know that wind and solar energy will not appreciably affect their sales of natural gas, oil, and gasoline sales over the next ten to twenty years? Will the availability of biofuel additives for gasoline prolong our use of gasoline? Why does the BP advertisement make no mention of the word "nuclear"? Has BP considered if nuclear energy is vigorously pursued, it might significantly diminish their sale of gasoline and oil products?

Gaining public acceptability is where we need to start if we want nuclear electric power to be a large contributor in our future energy mix. Without public acceptance, making nuclear energy a reality will be difficult. We have to understand the benefits and the challenges that nuclear electric power offers. In addition, we need to better understand the entire picture of our energy challenges. The challenges include the threat of climate change, economic implications, and our need for continued national security. Our national security requires that we have secure energy resources. Energy independence would place us in a strong position of energy security and would strengthen our national security.

Public acceptance built from a better understanding of nuclear power doesn't mean just a better understanding of fission or capacity factors; it is the sum total: energy demand, the needed energy supply, and the needed reductions in carbon dioxide emissions, the potential growth in high-paying and highly-skilled jobs, and the revitalization of heavy manufacturing and construction.

Beyond public acceptability and the financial resources, how will we reestablish the nuclear research and development infrastructure to support nuclear growth? If the resurgence does take hold and our efforts result in new reactor orders, the need for continued research and development efforts will be even greater. It will be the continued advancements on nuclear technology that can provide assurance that the growth of nuclear energy can be sustained into the future.

Nuclear power plants can produce large amounts of electrical power in the range of 1,500 megawatts for each plant. Plans for construction of new nuclear electric power plants should be started in 2011. Nuclear power plants can easily provide between 50 percent and 80 percent of the 175,000 megawatts of the electrical power needed by the year 2020. Additional nuclear power plants can easily provide between 50 percent and 80 percent of the 175,000 megawatts of power needed by the year 2030. If the DOE nuclear hydrogen facility proves successful in FY 2015, construction of nuclear hydrogen plants can be initiated. The nuclear hydrogen power plant is designed to more economically meet the future projected demand for hydrogen during the years 2020 to 2030. The advantages of nuclear power are:

- Nuclear power plants are producing electricity at a price about one tenth the price of electricity produced by any other type of electric plant.

- Nuclear electric power plants are currently the only plants capable of economically producing the large quantities of electricity and hydrogen required to support an electric/hydrogen economy.

- Only nuclear electric/hydrogen plants can offer our automakers the opportunity to lead the world in the production and sale of electric/hydrogen/ hybrid-powered vehicles.

- Nuclear electric power plants will allow a marked increase in the number and operation of electric/hydrogen-powered ground vehicles. Non-polluting vehicles fueled with stored electric energy and hydrogen will result in the virtual elimination of fossil-fueled vehicles and their associated exhaust pollutants.

- Nuclear electric plants provide "green power" and produce no adverse air pollution.

- [US] Our nuclear plants are producing rated power during 90 percent of the year, far higher than the power produced from other fossil fueled, solar, or wind power plants.

The Generation IV Nuclear Energy Systems Initiative within the DOE's Office of Nuclear Energy constitutes our contribution to an international effort to develop next-generation nuclear energy technologies.[55] The DOE received an appropriation of $179 million in FY 2009 and $220 million in FY 2010 for the Generation IV Nuclear Energy Systems program.

Whether we build new plants or not, other nations are building them now and they will be building more in the future. Recent international events and proliferation concerns have underscored the need to reexamine our policy and practices. Rising energy demands, our security, our prosperity, and our environment all require reducing our dependence on fossil fuels that emit greenhouse gases. No serious person can look at the challenge of maintaining our national security, reducing greenhouse gases, and addressing climate change and not come to the conclusion that nuclear power has to play a significant and growing role. To foster that growing role, our nuclear energy policy itself must take on a more significant role to be technologically robust, economically sound, and publicly acceptable for decades to come.

Unfortunately, the Obama administration appears to be much less enthusiastic about hydrogen than the Bush administration. The Obama administration argues that a variety of technical and market barriers make it unlikely that hydrogen will come into widespread use as an energy carrier. For FY 2011, the administration's $137 million request for hydrogen RD&D represents a $43 million cut from the FY 2010 appropriation, taking the program below the funding level it had in FY 2004.

Here we have another example of how a new administration can change the course of earlier plans that were designed to reach our goal of energy independence. Do you suppose the Obama administration has been listening to proponents that favor the expansion of our biofuel industry? Remember, the DOE proposed in 2006 a practical scenario for the production and distribution of hydrogen, when and if a Washington administration decided to promote the use of it.

Nuclear Waste Project Impedes Energy Independence

Since 1982, the DOE has been funding the design and construction for an underground storage facility for nuclear waste at the Yucca Mountain site near Las Vegas, Nevada. Under the 1982 Nuclear Waste Policy Act, the federal government is required to provide a method to safely store the spent fuel rods removed from our nuclear electric power plants. Our government is committed to provide a nuclear waste storage facility so that nuclear reactor owners are not saddled with the costs and responsibilities of managing and safeguarding the waste.

The DOE was scheduled to complete the storage facility by January of 1998. At the end of 2000, the total expenditures for the project reached about $58 billion. The nation's consumers have been paying one-tenth of a cent per kilowatt-hour of electricity produced by our nuclear plants.

In mid 2009, there were about 1,400 people working on the project. Because of congressional funding reductions for the project, the number diminished to 600 employees in 2010. President Obama's administration decided to eliminate all work on the Yucca Mountain Project in February of 2010. Funding is eliminated for the project in FY 2011; however, the administration increased funding to $54.5 billion for taxpayer-backed loan guarantees for new nuclear reactors.

In February of 2010, the DOE established the Blue Ribbon Commission on America's Nuclear Future to conduct a comprehensive review of nuclear waste disposal policies and to recommend alternatives.

Natural Gas-Fired Electric Power Plants Will Expedite Energy Independence

The US Energy Information Administration reports that our onshore supply of natural gas is projected to provide about 80 percent or more of the increase in domestic production anticipated through 2030. It is apparent that our oil companies do not have an ample current or future supply of natural gas to meet the demands. In order to have an ample supply of natural gas, a limited amount of new exploration is needed. For exploration to take place, our oil companies need access to the areas, offshore and onshore, that are known to have the potential to produce the natural gas needed to meet consumer demand. We need sufficient amounts of natural gas to operate new electric plants. The government must allow our oil companies access to a number of onshore and offshore fields having high potential for production of natural gas.

New natural gas-fired electric power plants must be constructed and placed on line during the next twenty years. Consumption of natural gas by residential and commercial consumers can be decreased by increasing our energy efficiency. Residential and commercial consumption can be diminished by switching from gas to electricity as new electric plants become operational. By decreasing residential and commercial demands for natural gas, the surplus can be used to supply new gas-fired electric plants.

Geothermal Power Plants Will Expedite Energy Independence

Geothermal is a "green power" resource that is limited, but it can provide a substantial amount of base-load and peak electric power needed. It was found entirely feasible to operate a geothermal plant for extended periods of time with no need for on-site personnel. The ability to operate a plant with little or no need for on-site personnel has important economic implications for commercializing hot dry rock technology. It should be selected as one of our resources for producing electric power.

It is concluded that geothermal power plants can provide a significant share of the 350,000 megawatts of electric power needed by the year 2030. With a cumulative number of possible geothermal power plants in operation, between 20,000 and 75,000 megawatts of electrical power can be provided. Design and construction of new geothermal plants should be initiated this year. As recommended by the DOE/EERE, technology for development of new sites and drilling techniques are needed. However, these efforts should be accelerated and funded to obtain high levels of geothermal energy that are estimated at 100,000 megawatts of electric power.

Modernized Transmission of Electricity Will Expedite Energy Independence

The DOE Office of Electricity Delivery and Energy Reliability is helping to modernize our nation's infrastructure. The nation's ability to meet the growing demand for reliable electricity is challenged by an aging electricity transmission and distribution system and by the vulnerabilities in our country's energy supply chain. Despite increasing demand, we have experienced a long period of underinvestment in power generation and infrastructure maintenance. The majority of the power delivery system was built on the technology developed in the 1960s, '70s, and '80s, and it is limited by the speed with which it can respond to disturbances. The limitation of response increases the vulnerability of the power system to outages that can spread quickly and have regional effects. Deploying the next generation of clean energy sources will require a complete modernization of our electric energy transmission infrastructure, which will rely on digital network controls and transmission, distribution, and storage breakthroughs.

The DOE Office of Electricity Delivery and Energy Reliability received an appropriation of $172 million for FY 2010 and requested an appropriation of $186 million for FY 2011, as reported in their FY 2011 Congressional Budget Request. Their allocation for FY 2009 was $4.5 billion, which was allocated from the American Recovery and Reinvestment Act, 2009, for modernizing and securing the electric grid.

Of the $186 million FY 2011 request, $144 million will be used for research and development to support the technologies that will improve the reliability, efficiency, flexibility, functionality, and security of the nation's electricity delivery system. The funds will also be used to develop transmission technologies and energy storage to enable more efficient integration of variable generation. Support is included for the next generation of smart grid technologies and for cyber security to ensure the increasingly digitized electrical infrastructure can be protected from cyber attacks

Coal-Fired Electric Power Plants Will Impede Energy Independence

The DOE was allocated about $480 million in fiscal year 2008 for research and development activities related to Coal. In fiscal year 2009 the DOE had funding appropriations of about $690 million for their Coal program. It is likely that a portion of the $3.4 billion received for fossil fuel R&D from the FY 2009 American Recovery and Reinvestment Act will be allocated for coal R&D.

Why did these funding allocations increase so much in 2009? Is it because our government wants to keep and expand the coal companies? We know our railway corporations want to continue carrying tons of coal. If we can use more "clean" coal our electric utility companies won't have to worry about using those pesky, high priced, renewable energy sources. How about the auto and natural gas industry? If we can burn more "clean" coal to produce electricity, the auto industry can continue to produce air polluting gasoline engines and the gas company can sell more natural gas for gas-fired electric plants.

The DOE provides a list of the "benefits" for carbon dioxide capture and underground storage.

- Coal, together with carbon dioxide capture and geologic storage" (CCS), allows our country to obtain economic benefits and energy security from our large domestic coal resources while under significant carbon dioxide emission constraints.

- Coal/CCS is not currently cost effective. Most cost-reduction opportunities are in the area of carbon dioxide capture.

- Barriers to carbon dioxide storage include safety, permanence, and geological storage capacity. Members of the Regional Carbon Sequestration Partnership are beginning to implement large-scale carbon dioxide storage tests in locations throughout our country and in Canada.

A significant number of additional demonstration projects carried out under the clean coal power initiative program are intended to prove the commercial viability of a suite of coal/CCS technology options.

- Widespread commercial deployment will require an extensive carbon dioxide transportation infrastructure, indemnification framework, regulatory certainty, and public acceptance.
- Coal/CCS may be ready for mass commercial deployment in selected applications by 2020.

The DOE list of "benefits" should have the heading "barriers" to inhibit further funding and efforts directed for the capture and underground storage of carbon dioxide.

- It is true; we do have enough coal in our country to last for 2,000 years. Certainly coal reserves will be a benefit if we can use coal for clean production of energy. But why continue to spend billions of dollars to learn how, when we have other more economical and available sources for producing clean energy?
- Second, on the DOE list, coal/CCS is not an economical option.
- Third, DOE admits that barriers to carbon dioxide underground storage include the issue of human safety. Remember Lake Nyos in Cameroon?
- The DOE has identified that the longevity of underground storage of carbon dioxide is an unknown barrier.
- Geological underground locations present another unknown with regard to the capacity for carbon dioxide storage. The DOE plan to implement large-scale underground storage of carbon dioxide in locations throughout our country is unconscionable and should be stopped in its tracks. The taxpayers should

not have to pay for cleaning up carbon dioxide storage sites in the future if and when a site is found to be leaking carbon dioxide into the atmosphere.

- The fifth item on the DOE "benefit" list reports that commercial carbon dioxide sequestration will require an "extensive transportation infrastructure." In other words, long pipelines, which may leak, are needed to transport carbon dioxide across the countryside. Where new pipelines are installed, environmental regulations must be met, as well state and landowner approvals. Other barriers to a pipeline system include government compensation for damage, loss, or injury suffered. The DOE also points out that the government must, through public education, gain public acceptance of this initiative.

- The last "barrier" on the DOE "tax payer benefits" list is the revelation that we, the taxpayers, need to continue to pay for research, development, and testing for the next ten years. If the DOE continues to spend our money for the next ten years at previous rates that have averaged about $500 million each year, the taxpayer will be expected to pay a total of about $5 billion. Adding on the $3.4 billion allocated to coal from the American Recovery and Reinvestment Act, 2009, this brings the taxpayers total bill to $8.4 billion. The DOE tells us: "We may be ready for commercial deployment in selected underground sites by 2020".

Biofuels Will Impede Energy Independence

A tidy portion of the money made available by The American Recovery and Reinvestment Act of 2009 is being used to promote biofuel production and consumption.

- President Obama announced in May of 2009 that the DOE plans to invest $786.5 million to accelerate the advance of biofuel research and development and to provide additional funding for bio-refinery demonstration projects. He also announced that his administration is taking several steps to advance biofuel research and commercialization. The steps are planned to help preserve biofuel industry jobs and to establish a Biofuels Interagency Working Group. This group will have the assignment to develop the nation's first comprehensive program for growing the biofuels market.

- In December 2009, the Department of Energy announced the selection of nineteen integrated bio-refinery projects to receive up to $564 million to accelerate the construction of commercial-scale facilities. The projects, located in fifteen states, will be used to validate refining technologies and help lay the foundation for full commercial-scale development of a biomass industry in our country. The projects selected will produce advanced biofuels, bio-power, and bio-products using biomass feed stocks at the pilot production plant size, demonstration plant size, and full-scale commercial plant size. These projects will be matched with more than $700 million in private and non-federal cost-share funds. The total project investment is expected to total almost $1.3 billion.

The fact is that construction of nuclear and gas-fired electric power plants and the required electric transmission infrastructure can start now. These plants can produce the electricity and hydrogen that is needed once they are up and running. The required distribution infrastructure for hydrogen can start now. Why continue to speculate on the possible future success of biofuels that need production plants and a distribution infrastructure? A commitment for electricity and hydrogen as the primary source of energy can start today.

How many gallons of biofuels do we consume? Since 2001, ethanol production has increased from 1.6 billion gallons in 2000 to an estimated 6.4 billion gallons in 2007. The vast majority of the ethanol was produced from corn. In 2007 we produced about 450 million gallons of biodiesel fuel. In 2005, our country became the world's leading producer of ethanol. By 2007 we accounted for nearly half of worldwide ethanol production. The Energy Independence and Security Act 2007 sets a mandatory renewable fuel standard requiring fuel producers to use at least 36 billion gallons of biofuels in 2022.

How many dollars is our government spending for biofuels? The high cost, compared to gasoline, of producing and transporting ethanol will continue to limit its use as a renewable fuel. Ethanol relies heavily on federal and state subsidies, enabling it to remain economically viable as a gasoline blending component. The federal tax credit of 45 cents per gallon makes it possible for ethanol to compete as a gasoline additive.

We gave ethanol producers about $3.0 billion and biodiesel producers $180 million in the form of federal renewable energy tax credits during 2007. The federal bill for ethanol increases with every additional gallon of ethanol produced. It is estimated that during the year 2010, ethanol will have cost the taxpayers more than $5 billion.

What can we make of this initiative? For one, three quarters of a billion dollars for this project is only the tip of the iceberg. Why do we need a new Biofuels Interagency Working Group? Will the working group figure out where to spend the next multi-billion dollar chunk of our tax money for biofuels? Part of the reason behind the congressional formation of the Department of Energy was to "combine the previous separate functions under one department to improve coordination analysis of costs and benefits." Apparently the biofuels project is so large that the DOE can't manage it alone, and they need help from the Biofuels Interagency Working Group. A portion of the $700 million in private and non-federal cost share funding may be supported by our oil companies.

Another reason for spending more money on biofuels, according to President Obama, is to preserve biofuel industry jobs. The Works

Program Administration 1935 (WPA) provided federal funding during the late 1930s to keep people working on jobs during the Great Depression. Some people agreed that many of the projects funded by the WPA were not needed. Is it possible that federal funding used to support the biofuel industry like the WPA may not be needed? Maybe the only purpose for increasing the size of the biofuel industry is to provide jobs for the unemployed.

If biofuels are selected now to help bridge the time between now and the year 2022, hundreds of biofuel production plants are needed and a huge distribution infrastructure is required. Why should we invest in an infrastructure to support biofuels when hydrogen will eventually be required to replace fossil fuels and biofuels? A time period for the introduction of hydrogen as a replacement for gasoline has not been established. No firm time lines have been established to implement a plan to build the infrastructure needed for hydrogen production and distribution. Some DOE reports estimate the phase in time period for hydrogen may occur around the year 2050. By the year 2030, an effort will be required to enable the use of hydrogen by the year 2050. Hundreds of electric power plants will be required with facilities for the production of hydrogen. A distribution infrastructure will be required for hydrogen and electric power. Eventually all of our billions of tax dollars invested in the biofuel industry will be lost. With the eventual introduction of the electric-hydrogen economy, biofuel production plants and the transportation infrastructure will lay idle.

Wind Energy Will Impede Energy Independence

Some of the disadvantages of relying on wind turbines for producing electric energy are:

- They are unreliable for supplying base-load electrical power and for supplying peaking power.

- The cost of electric energy produced by wind turbines exceeds the average consumer electric rate of 6 to 7 cents per kilowatt-hour.

- Wind-turbine electric energy does not reduce fossil fuel imports and, as a result, will not lead to energy independence.

- Taxpayer money is being used to subsidize private companies to invest in and install wind turbines. Some of our tax money is used to purchase wind turbine components and completed wind turbines from foreign companies. United States Senator Sherrod Brown, from Ohio, criticized a $450 million federal stimulus grant to a Texas wind farm that will be built with Chinese wind turbine components.[56] The senator said the project would generate 3,000 jobs in China and 330 jobs in our country. Of the 330 jobs in our country, 300 would be temporary. On the other hand, Stephanie Mueller, spokeswoman for the Department of Energy, said the American Reinvestment and Recovery Act 2009 has attracted more than $10 billion in foreign investment. The investment has created wind-power jobs in our country.

- Energy storage will increase the already-proven high cost of energy produced by wind turbines.

- Large wind turbine farms located at remote onshore wind sites require new electric lines to connect with existing utility networks. New electric power lines and sophisticated power control systems are needed and will add another increment of cost for the energy produced by wind turbines.

Offshore wind turbine farms, located in the ocean or in the great lakes, pose the same disadvantages as the land-based wind turbines. However, offshore wind turbines have the additional disadvantages as listed below.

- Each wind turbine must be built to protect against deterioration from offshore environments. The cost to build a wind turbine for offshore operation is more than the cost of a comparably sized wind turbine built to operate on dry land.

- Foundations for wind turbines sited in the water cost more than foundations for a comparably sized wind turbine sited onshore.

- The cost to operate and maintain a wind turbine sited in the water is more expensive than the cost to operate and maintain a comparably sized wind turbine sited onshore. Expenses required for the operation and maintenance of a wind turbine directly affects the cost of energy that it produces.

- Offshore wind turbine farms will require new electric lines both underwater and on land to connect with existing utility networks.

- The cost of energy from offshore wind turbines is easily twice the cost of energy produced by onshore wind turbines.

- Federal subsidies will be needed to stimulate private investors to acquire and install wind turbines for offshore energy production. Offshore wind turbine farms will pose higher investment risks compared to the investment risks presented for land-based wind farms. As a result, private investors will require, as an incentive, higher federal subsidies than the 2009 subsidies for land-based wind farms before investing in offshore wind turbine farms.

- Wind turbine farms sited in oceans and lakes pose potential risks to marine life environments, private boater safety, and the safe operation of commercial lake and ocean shipping.

The city of Berea, Ohio, plans to install a refurbished German wind turbine at the Cuyahoga County Fairgrounds. County officials estimate the turbine will generate enough power to cut $50,000 from the fairgrounds' $90,000 annual energy bill. Keep in mind, the fairgrounds, with their wind turbine, can save $50,000 each year. Cuyahoga County is receiving $1,000,000, from the Recovery and Reinvestment Act 2009 to purchase the refurbished German-made wind turbine. The county is receiving an additional $400,000 grant from the DOE Energy Efficiency Program to help with the wind turbine project. The county has applied to the state of Ohio for an additional $200,000 grant that will be used to help pay for and install the wind turbine. To summarize the deal, the taxpayer is spending $1.6 million in order to save the fairgrounds an estimated $50,000 each year.

Why should we pay for the purchase and installation of a refurbished German wind turbine? A financial consultant might suggest that our government not pay for the refurbished wind turbine but invest the $1.6 million at an annual rate of 3.1 percent. The investment would yield $50,000 each year. The annual $50,000 yield could be made payable to the Cuyahoga County Fairgrounds. The fairgrounds could save additional money by eliminating their expense for operating and maintaining the wind turbine and eliminate their expense for its eventual removal.

Tidal and Wave Action Energy are an Impediment

Some of the disadvantages of tidal and wave action as a renewable energy source are:

- Tidal and wave action power plants are in their early stages of development. As a result, the cost of energy produced from tidal power or wave action is not competitive with conventional fossil-fueled or nuclear power plants.

- Wave action energy production can conflict with merchant shipping, recreational boating, and commercial fishing activities.

- Devices for the production of energy from wave action, including the oscillating water column and the point absorber, are in their early stages of development. It is highly questionable as to whether or not these devices will ever lead to large scale production of energy. These devices may have military applications where small amounts of energy are needed.

- Tidal energy sites are limited to the northeastern and northwestern parts of our country. These sites have a low potential for producing significant amounts of electric energy. For example, the tidal action plant planned for installation on the Cape Cod Canal is expected to produce 30 megawatts of electric power.

- Tidal energy plants produce energy during an ebb tide, which occurs about two times over a twenty-four-hour period. Like solar energy and wind energy, tidal energy plants are not able to provide baseline power or peaking power to our electric utility plants. Unless provisions for energy storage are built into each tidal energy plant, these plants can only act as fuel savers. Fuel-saver energy sources do not diminish our need for gasoline. They will not contribute to our energy independence.

In our country, ocean energy development has historically been stifled by a number of state and federal regulatory hurdles. These regulatory issues appear to be relaxing, but it may be years before regulations allow advanced development of tidal and wave action energy extraction.

It doesn't cost much to operate the tidal water power plants. But their construction costs are high, which lengthens the time required before a profit may be realized. As a result, the cost per kilowatt-hour of electric energy derived from tidal power or wave action is not competitive with conventional fossil-fueled or nuclear electric power plants.

Solar Voltaic and Solar Thermal Energy Impedes Energy Independence

Solar Photovoltaic Arrays

- The cost of energy in 2009 is over three times the cost of energy being produced by our electric power plants.

- With solar photovoltaic efficiencies of about 10 percent, as much as 30,000 square miles of land is needed to enable solar plants to supply our energy demands.

- Installation of large photovoltaic arrays must be located in the southwest regions of our country. A method is needed to transport energy from the southwest to the north-eastern and northern mid-west locations where large amounts of electrical energy is consumed.

- Solar photovoltaic arrays cannot be relied on to supply base-load power or peak electric demands.

- A method for storing solar energy is needed. Energy storage will add to the basic high cost for solar energy. It is estimated that facilities acquired for and prepared to accommodate compressed air storage would add at least 3 to 4 cents per kilowatt-hour to the cost for photovoltaic electric generation.

- Some solar advocates propose building a new electrical grid infrastructure to transport high-voltage direct current electricity from the southwest to the eastern energy consumers. Our existing infrastructure for transporting high-voltage alternating current electricity cannot be used to transport direct current electricity produced by solar photovoltaic arrays. The investment needed for a new high-voltage direct current infrastructure will add another increment to the already high cost of electricity produced by photovoltaics.

Solar Thermal Concentrators

Solar Thermal Concentrators, introduced earlier, include the linear concentrator, the collector dish/engine system and the power tower system.

- For economical operation of solar thermal concentrators, they must be installed in the southwest locations of our country.

- Solar thermal linear concentrators require large tracts of land to produce the amount of electricity that we need. The Nevada Solar One, a solar plant located in boulder City, Nevada, produces a rated output of 64 megawatts. The plant employs 760 parabolic concentrators with more than 18,240 receiver tubes and occupies an area of 400 acres. For a similar plant rated at 1,500 megawatts the land area needed would be 9,375 acres or about 14.6 square miles. If the area used was contained with in a square, each side of the square would be 3.8 miles long.

- Solar concentrators are not able to supply our electricity infrastructure with base-load power or peak power demands.

- A method for storing thermal energy is needed. Thermal storage tanks are capable of storing thermal energy but add to the basic cost of energy produced.

- Collector dish and engine systems are generally used to produce electricity in the low kilowatt range of 3 to 25 kilowatts. A demonstration 10-megawatt solar-power tower system funded by the DOE successfully produced electricity, but it was shut down in 1995 because it was not commercially economical.

- The DOE plan for concentrated solar power in FY 2011 calls for developing low-cost systems with thermal storage to achieve cost competitiveness in the intermediate and base-load power markets. The DOE will assist industry deployment by identifying land environmentally suitable for utility-scale solar projects and address issues related to water consumption and transmission. The DOE will launch an initiative to demonstrate new concentrated solar power technologies that "could" lead to over 1 Gigawatt of projects.

- Instead of working on these technologies for ten years and continuing to spend hundreds of millions of our dollars each year, why not build, within two years, a gas-fired electric power plant rated at 1,000 megawatts? We know the gas-fired power plant will work. The DOE is not entirely sure that their solar power technologies will ever produce 1 Gigawatt.

Energy Conservation Not a Significant Driver for Energy Independence

Billions of our tax dollars are being spent for Weatherization and Intergovernmental Energy Efficiency and Conservation Block Grants.

The Weatherization and Intergovernmental Activities program is a new budget line item found in the FY 2011 Department of Energy Congressional Budget Request. The Weatherization Assistance Program, through a state-managed network of local weatherization providers, supports home-energy retrofits for low-income families and career development opportunities for workers. The innovations in weatherizing activity will continue to demonstrate new ways to increase the number of weatherized low-income homes and to lower the federal per-home cost for residential retrofits. Tribal Energy Activities supports feasibility assessments and project planning for clean energy projects on tribal lands.

The mission of the DOE industrial technologies program is to significantly reduce the intensity of energy used by our industrial sector through research, development and demonstration of next-generation manufacturing technologies. By reducing energy consumption associated with industrial processes the program will reduce our national dependence on foreign energy sources. The DOE estimates, as a result of this project, a cumulative reduction of at least 200 million barrels of oil imports by 2030 and ten times that amount by 2050. In other words from 2011 to the beginning of 2030, 19 years, the DOE estimates that we will have reduced our imports of oil by a total of 200 million barrels. On the average, over the 19 years, we will have saved about 10.5 million barrels of imported oil each year. In 2010, on average we consumed about 360 million barrels of imported oil each month or about 4.3 billion barrels for the year. Fifty five percent of our oil imports are from the Persian Gulf and OPEC nations. It is likely that our oil imports for 2011 will be about the same quantity of oil that was imported in 2010. The percent of saving, as a result of the industrial technology program, for 2011 is therefore expected to be about 0.24 percent of the total amount of imported oil in 2011. To continue to support this DOE research and development activity with our tax dollars, the reduction in imported oil does not appear to provide an impressive return on our investment.

The industrial technology program can easily stifle the free market system for industries critical to the nation's economic prosperity and national security. These industries must maintain their economic health and profit margins before implementing changes for the purpose of increasing the energy efficiency of their operation. Foreign competition and the free market system will be the deciding factor for energy efficiency improvements.

The Building and Energy Codes activity, managed by the DOE, provides technical and financial assistance to states to update, implement, and enforce their computerized energy codes to meet or exceed the computer model codes. The DOE is required to continue this activity by Section 304 of the Energy and Conservation Act, 1976, and the Energy Independence and Security Act, 2007. Is it time to re-examine the need for this legislation and examine the need for DOE to continue to update, implement, and enforce the codes? Why not have each state take on the responsibility for updating, implementing, and enforcing the building codes? On the other hand, the DOE should maintain the building energy code website. Users can download the compliance software and training materials free of charge. The DOE should continue to support a technical staff to answer questions on energy codes for buildings and connect inquires to other resources.

Recall the Works Program Administration, 1935

The money allocated by the American Recovery and Reinvestment Act of 2009 is being used in ways that are similar to the Work Projects Administration (WPA) projects from 1935 to 1943. The WPA program supported the construction of public buildings, roads, and other projects. Almost every community in America has a park, bridge, or school constructed by workers that were employed and paid by the WPA. By 1938, about 3.3 million workers were employed by the WPA. Like most government programs, there were critics. One criticism of the program was that it wasted federal dollars on projects that were often not needed or wanted. The allocation of WPA projects and funding was often criticized because the projects were often made for political considerations. Politicians who were favored by the Roosevelt administration or who possessed considerable seniority and political power often helped decide which states and localities received the most funding.

The most serious political criticism was that President Roosevelt was building a nationwide voter base with the millions of people that were supported by the WPA funding. Some historians have concluded that the WPA projects did not strongly stimulate the depressed economy from 1935 to 1938. From 1939 to 1943, the economy and employment experienced a tremendous rebound. Because of the large number of men and women needed and employed to support the war effort Congress terminated the WPA in 1943.

Clearly we need to increase our energy efficiency in all end use sectors. However, even aggressive efforts to improve efficiency cannot eliminate replacement and new capacity additions needed to avoid severe reductions in the services that energy provides to all Americans.

Brad Linscott

The Common Sense Energy Plan

Introduction

Congress has taken a reactive approach to resolve several catastrophic events during the last ten years. In some instances, because of a perceived complacent posture, their inability to prevent these events has adversely affected a large number of our citizens. The effects include financial loss and loss of life for thousands of our citizens. Some examples sited include the credit union bailout, the attack on September 11, and the most recent financial institution bailout, including financial lending to US automakers. A proactive approach is one that anticipates future events and one that takes appropriate action designed to prevent adverse events from occurring. Congress needs to begin a new approach, look ahead to the future, anticipate problems, prepare proactive plans, and implement these plans to significantly reduce reactionary efforts in the future. By implementing the Common Sense Energy Plan, Congress can start a proactive approach to solve our energy problems before the occurrence of future adverse situations that require reactionary measures.

The majority of the people in our country are concerned about our budget deficit and the fact that our deficit is growing larger each day. Part of the solution for deficit reduction is to reduce our government expenditures and to reduce the number of federal employees. Can we reduce future budgets for the Department of Energy? In addition, can we save billions of dollars planned for future federal subsidies and tax incentives to support renewable energy? The Common Sense Energy Plan says yes, we can!

How much of our tax money is being spent by the DOE? The DOE requested $28.4 billion for FY 2011. The FY 2011 request increased by $1.8 billion compared to the DOE request for FY 2010. The FY 2011 funding request supports the president's commitment to respond in a considered, yet expeditious, manner to the challenges of rebuilding the economy, maintaining nuclear deterrence, securing nuclear materials, improving energy efficiency, incentivizing production of renewable energy, and curbing greenhouse gas emissions that contribute to climate change.[57] Together, the American Recovery and Reinvestment Act of 2009, the FY 2010 budget allocation, and the FY 2011 budget request will support investments for a multi-year effort to address these interconnected challenges.

By the end of FY 2010, the DOE will have obligated 100 percent of the $36.7 billion allocated by the Recovery Act 2009 and will have spent about 35 to 40 percent of the $36.7 billion. At the end of FY 2010, the DOE will have spent about $39.4 billion of our tax money.

The intent of this so-called "investment" is to strengthen our economy by providing much-needed investment by saving and creating tens of thousands of direct jobs, cutting carbon emissions, and reducing our dependence on foreign oil.

Of the total FY 2011 Department of Energy request for $28.4 billion, $2.3 billion will be allocated for the DOE's FY 2011 Energy Efficiency and Renewable Energy program. Andrew Grove, former chief executive of the Intel Corp, suggests that our country needs an overarching national goal to help us prioritize and evaluate the myriad of energy-related proposals and programs. We need an overarching energy plan that, when implemented, will provide economical supplies of energy and eliminate our dependence on imported oil, natural gas, and electricity. As discussed earlier, we do, in fact, import natural gas and electricity as well as oil.

If our government continues to directly fund and subsidize all of the energy related activities, as reported by annual Department of Energy Congressional budget requests, we will continue to import huge amounts of foreign oil. And in return we will continue to ship billions of dollars to the Middle Eastern countries. Alexander Karsner, former assistant secretary for energy efficiency and renewable energy, suggests, "Every bit of money that goes away is a lost opportunity for money we could have invested here."

Our country is focused on energy research and technology development. We are subsidizing a variety of renewable energy sources. Many of the DOE-funded programs including solar, wind, and biofuels, will continue each year in the future to ask for more of our tax money. Typically the cost of renewable energy is significantly higher than the cost of energy being produced today. We are told more tax money is needed to help find new ways to reduce the cost of renewable energy. These activities are planned to continue for many years in the future. During our future years we will see little progress toward reducing our foreign oil imports. If all of the Department of Energy R&D efforts are allowed to continue, they will only serve to delay a decision that is needed now. Congress needs to reduce or eliminate funding for some of the less promising energy related R&D projects. They need to select an energy plan that is supported by well-proven technology. By implementing such a plan, the energy produced will be clean and economical and will significantly reduce our need to import foreign oil.

Is the government's policy for renewable energy development going to speed us toward energy independence? Our energy policy will not allow us to reach energy independence for at least twenty years. Solar, biofuels, ocean wave, and tidal action are each in an "infant stage" of research and development. Each requires continued research, development, and testing. It will be at least ten to fifteen years before we can start construction and operation and begin to install the renewable energy distribution infrastructure needed. The time needed for this effort can easily exceed twenty years and require hundreds of billions of dollars of federal funds before they can produce the large amounts of renewable energy we need. Remember the DOE has spent over $39 billion in FY 2010. Do we want ten more years of yearly spending at the FY 2010 level?

Over 80 percent of the world's supply of petroleum is controlled by countries that have the power to manipulate our supply with relative impunity. Certain Middle Eastern nations that supply oil to the world believe it is their right to extort outrageous profits from their customers. Unfortunately for us, some of their profits are used to support anti-American terrorist organizations. For this reason alone, it is understandable the American public believes the right thing to do is to demand that our government take action to insure the development and availability of energy sources that will force a return to a balance of global energy independence.

The Common Sense Energy Plan, when implemented, will reduce our energy imports within ten years. The five objectives of the Common Sense Energy Plan are to:

1. Provide energy independence

2. Reduce federal spending for selected energy options

3. Reduce air, land, and water pollution

4. Reduce consumption of fossil fuels.

5. Organize to implement the plan

1.0 Provide Energy Independence

1.1 Build Nuclear Electric Power Plants

The Common Sense Energy Plan calls for using our resources to move toward an electric/hydrogen economy. The DOE has reported a feasible scenario that supports the hydrogen economy that could start today. More nuclear electric plants must be built and put into operation. They can produce huge amounts of electricity with no emissions of carbon dioxide. New nuclear electric plants being developed by the DOE can consume, as fuel, large inventories of radioactive waste elements that have been generated by our operating nuclear electric plants since 1960. The new nuclear reactors, in the process of producing energy, destroy a great proportion of the long-lived radiotoxic constituents, where otherwise the radiotoxic constituents would require isolation in a geological-repository. New nuclear plants are also being developed by the DOE to efficiently produce hydrogen.

The DOE received an appropriation of $791 million for the technology development of non-defense nuclear energy in FY 2009. $870 million was appropriated in FY 2010 and $912 million is being requested for FY 2011.

No funding was allocated from the American Recovery and Reinvestment Act 2009 for the DOE non-defense nuclear energy efforts. But $16.8 billion was allocated to the DOE in 2009, by the Recovery Act, for Energy Efficiency and Renewable Energy.

How many new power plants do we need and when? We need about 150 new nuclear power plants that produce on average 1,500 megawatts each. By 2020 we need 75 of the 150 nuclear electric plants on line. By 2030, we need the remaining 75 plants producing a mix of electricity and hydrogen.

A total of 150 new nuclear plants may sound like an enormous number to have operating in our country. France, for example, developed nuclear technology and made it a centerpiece of their energy portfolio. Eighty percent of its power is nuclear-generated. The entire country of France is 50,000 square miles. France is smaller than our state of Texas, and yet they have fifty-nine nuclear power plants currently on line. Japan, about half the geographic size of France, has fifty-five nuclear power plants on line. We have 104 nuclear power plants that produce 20 percent of our electricity. If 150 new nuclear power plants were on line by the year 2030, and the plants were geographically divided equally to each of our fifty states, each state would have only 3 new nuclear plants on line.

1.2 Build Natural Gas-Fired Electric Plants

New natural gas-fired combined-cycle electric plants are needed that use high-efficiency steam and gas turbines. The Common Sense Energy Plan calls for building about 150 new natural-gas-fired combined-cycle electric power plants having an overall average rating of 1,500 megawatts between 2011 and 2031.

The generation of carbon dioxide resulting from natural gas combustion is about 50 percent less than carbon dioxide generated from coal combustion. The new gas-fired combined-cycle power plants are reaching thermal efficiencies that are almost double the thermal efficiency of the coal-fired steam boiler plants. The combined-cycle plant uses a combination of steam turbines and gas turbines to drive electric generators. This means the new gas-fired plants are able to produce more electricity with less natural gas than the gas-fired steam boiler plants. By using less natural gas, a further reduction of carbon dioxide is realized when compared to a coal-fired steam power plant.

The natural gas industry is not allowed, by our government and some states, to install new offshore drilling rigs. To maintain adequate supplies of natural gas in our country, we are importing natural gas from other countries. The gas industry is drilling for, and extracting natural gas from, new underground locations across our country. In certain instances, gas drilling in Pennsylvania, New York, and Ohio has caused the pollution of freshwater streams and residential water supplies. To maintain adequate supplies of natural gas, Alaskan and offshore drilling for natural gas is needed.

To conserve and to assure adequate natural gas supplies for the new power plants, a reduction in natural gas consumption by residential customers is needed. With the increase in electricity provided by new nuclear, natural-gas, and geothermal power plants, the residential customers can reduce their natural gas consumption by heating with electric heat pumps. For cold climate conditions in our country, where the efficiency of a heat pump is degraded due to the cold, hydrogen gas can be used as a fuel for heating instead of fuel oil or natural gas.

1.3 Build Geothermal Energy Plants

Considering all of the available renewable energy resources, geothermal electric power plants should be designed, for suitable site locations, installed, and put into operation. The technology and demonstration of enhanced geothermal systems using hot dry rock resources was successfully proven during tests by the scientists and engineers at the Los Alamos National Laboratory between 1970 and 1991.

The DOE increased their funding for research, development, and demonstration of geothermal energy production since their appropriation of $19 million for FY 2008.

For FY 2009 and FY 2010, the DOE received an appropriation of about $43 million and $44 million respectively. The American Recovery and Reinvestment Act, 2009, allocated $393 million for geothermal technology. The DOE requested $55 million for FY 2011.

To meet our future energy demands, installation and operation of geothermal electric power plants has to be accelerated. The expansion of a geothermal program may require federal subsidies, tax incentives, and additional DOE funding during the early phases. The Common Sense Energy Plan calls for construction of a sufficient number of geothermal electric power plants capable of producing a total rated power between 50,000 megawatts and 100,000 megawatts by 2030.

1.4 Modernize Our Electric Transmission System

With the plan to build new electric power plants, an upgraded electric transmission system is required. The largest consumers of electricity will be the large cities, their adjoining suburbs, and industrial sites. The federal government must provide the leadership and financial backing to complete a modified or, in some cases, a new network for transmitting a much larger amount of electric power.

The DOE Office of Electricity Delivery and Energy Reliability leads a national effort to modernize the electric grid, enhance security and reliability of the energy infrastructure and to facilitate recovery from disruptions to energy supply. The Research and Development program consists of four subprograms, Clean Energy Transmission and Reliability, Smart Grid Research and Development, Energy Storage and Cyber Security for Energy Delivery Systems. The DOE is requesting about $125 million for the FY 2011 Research and Development program. Two activities in the R&D program that pertain to renewable energy support should be eliminated. The subprogram Clean Energy Transmission and Reliability and the subprogram Energy Storage each directly support both wind turbine and solar renewable energy sources. Both subprograms should be eliminated and will result in a reduction in funding for this DOE R&D program.

1.5 Start Hydrogen Production, Storage, and Infrastructure for Distribution

The DOE has developed a scenario for hydrogen technology readiness to be complete by 2015. The goal is technology readiness of hydrogen production, delivery, and storage. The technology would enable automobile and energy companies to opt for commercially available hydrogen-fueled vehicles and a hydrogen fuel infrastructure by 2020. Our government needs to adopt hydrogen technology and enact policies to nurture the development of an industry capable of delivering significant quantities of hydrogen to the market place. Industry's role would become increasingly dominant as the market penetration of hydrogen increases into the energy market. Fully developed markets and infrastructure, transportation systems, and hydrogen storage and distribution can be commercially available across our country by 2040.

The Department of Energy is developing an advanced high-temperature nuclear reactor, referred to as a "demonstration plant," that will convert water into hydrogen and oxygen by 2015. The demonstration plant will be smaller in size than a commercial plant and will simulate and establish the viability of a larger commercial-sized plant. The DOE should be fully funding this project and planning for and, in the near term, implementing construction of a number of commercial-sized plants while the demonstration plant project progresses.

In FY 2008 and FY 2009, the DOE focused a portion of their research and development efforts on hydrogen storage. Their focus was to develop on-board vehicular storage systems that allow for a driving range of more than 300 miles. The effort was directed toward full market penetration across the North American light-duty vehicle market, within the constraints of weight, volume, safety, durability, refueling time, efficiency, and total cost to meet consumer expectations. The hydrogen storage activities concentrated on low-pressure, materials-based technologies. The project also explored advanced conformable and low-cost tank technologies for hydrogen storage systems to meet performance targets. The DOE effort on hydrogen storage coincides with the earlier discussion on the Toyota Prius automobile.

In FY 2008 and FY 2009, about $100 million was invested on the hydrogen storage project. Unfortunately, the project was not funded for FY 2010. Deferral of the effort in FY 2010 is aligned with DOE's portfolio of technologies for near-term impact, improved energy efficiency using multiple fuels, and job creation, consistent with presidential objectives. Funding for hydrogen storage is included in the FY 2011 budget, with fuel-cell technologies for a total request of $137 million. Hydrogen storage is an extremely important requirement to expedite the introduction of the electric/hydrogen economy. FY 2011 funding for fuel-cell technology should be eliminated and the amount saved added to the effort for hydrogen storage.

A transition to hydrogen will greatly improve our national security because it will eliminate our need for foreign oil. The movement to hydrogen will provide the opportunity for our automobile and truck industries to recapture market share lost to foreign multinationals during the last ten years.

1.6 Economic Prosperity

The transition from our fossil fuel economy to the electric/hydrogen economy will create millions of green jobs and increase the competitiveness of private industry and technology. Design and construction of electric power plant projects require private sector employment. Emerging hydrogen gas production, distribution, and end-use technology industries all promise new green employment opportunities. The change from fossil fuel to the electric/hydrogen option offers our automotive industries the opportunity to lead the world again in automobile industry employment, production, and sales.

Lower carbon dioxide emissions and clean energy will position our country to lead the world on climate-change policy, technology, and science. The electric/hydrogen program will leverage both domestic and international R&D partnerships to advance hydrogen development, which is aimed at demonstrating a viable pathway to support private sector deployment of hydrogen gas. The success of the electric/hydrogen program has clear international implications, as do partnerships with private and non-profit entities having influence that extends beyond the borders of our country.

Brad Linscott

2.0 Reduce Federal Spending for Selected Energy Options

Introduction

Having selected the electric/hydrogen economy, many government-funded programs can be reduced in scope or eliminated. For example, termination of funding for biofuel development can provide substantial savings. As an alternative, the funding planned for biofuels can be used to further support the electric/hydrogen effort.

To stimulate a more rapid change from our fossil fuel economy to the electric/hydrogen economy, government leadership is needed to plan, organize, and gain the popular support from the public. Funding authorization is needed with a congressional resolve to initiate and implement a plan supporting our need for an electric/hydrogen economy.

2.1 Wind Energy

The Common Sense Energy Plan calls for the elimination of federal subsidies and tax incentives for megawatt-sized wind turbines.

The federal tax credit for wind turbines installed during 2009 is estimated to have generated a loss of $1.4 billion in federal revenue. It is likely that the same or greater amount of revenue loss will occur in 2011.

The DOE appropriation for FY 2010 is $80 million. The American Recovery and Reinvestment Act 2009 allocated about $107 million in FY 2009 for wind energy. The DOE budget request for FY 2011 to support wind energy research and development is about $122 million. In FY 2011, the DOE is launching a robust offshore wind R&D effort to address technical, scaling, environmental, regulatory, and public acceptance risks to accelerate clean energy contributions from the nation's untapped offshore resources.

Do we as a nation want to spend more of our tax money for offshore wind turbines? Remember the cost of energy for offshore wind turbines will be twice the cost of energy for onshore wind turbines. About $1.5 billion can be saved in FY 2011 by eliminating all federal support for wind energy.

2.2 Solar Energy

The conversion of solar energy to electrical energy using state-of-the-art photovoltaic cells remains inefficient and too costly to build large, megawatt-size solar electric plants. Large solar electric plants should not be given serious consideration for use as a viable renewable energy resource in the near term. The cost of energy produced by solar electric plants is not economically competitive with our mix of coal, natural gas, and nuclear electric plants.

The largest hurdle for solar energy is to be able to significantly reduce the current cost of solar energy so that it will be competitive with today's energy cost. Some researchers at Rice University in Houston, Texas, advocate more funding support for alternative energy research. Some believe that unless research progresses far more rapidly to solve the current alternative energy problems by the year 2020, there is essentially no way to have large amounts of solar energy plants in place by 2050. The gap between solar energy's potential and what is needed to be practical on a massive scale remains wide.

The Department of Energy (DOE) funding for solar energy research and development has increased over the last few years. In FY 2008, the DOE received an appropriation of about $166 million and $172 million in FY 2009. Under the American Reinvestment and Recovery Act 2009, the DOE received an additional appropriation for solar energy of $116 million in FY 2009. For FY 2010, the DOE has an appropriation of $247 million, and for FY 2011, the DOE has requested a congressional appropriation of $302 million.

In three years, from the beginning of FY 2008 to the end of FY 2010, the DOE received authorization to spend about $700 million. With the requested appropriation of $302 million for FY2011, the total in four years, from 2008 to 2011, will amount to about $1 billion.

The Common Sense Energy Plan calls for the elimination of federal subsidies and tax incentives for design, fabrication, and installation of solar energy electric plants.

More than $300 million can be saved in FY 2011 by eliminating all federal support for solar energy.

2.3 Biofuels

The Department of Energy continues to conduct research and development of biomass and biorefinery systems. $217 million was authorized in 2009 for research and development related to biofuels and in 2010 the DOE requested $235 million. Biofuel research, development and production are continuing to assess three basic biomass materials.

The DOE funding request for feedstock infrastructure for FY 2010 is $12 million more than the money authorized for 2009. The effort will support increasing the number of production trials on various types of feedstocks. The additional production trials are critically needed to ensure a stable supply of feedstocks needed for a viable domestic biofuel industry. A biofuel industry goal is to provide sufficient production of biofuels to provide the volumes mandated by the Energy Independence and Security Act 2007. The act sets a mandatory Renewable Fuel Standard requiring fuel producers to use at least 36 billion gallons of biofuel in 2022. With a decision favoring the electric/hydrogen economy, the portion of the Energy Independence and Security Act 2007 related to biofuels can be rescinded.

Biomass and biorefinery systems research and development was increased by about $6 million for FY 2010, compared to FY 2009, to support ongoing multi-year biorefinery project deployment schedules. In FY 2009, the DOE was allocated a total of $777 million from the American Reinvestment and Recovery Act 2009 for biomass and bio-refinery systems research and development. Federal subsidies for biofuel production is estimated at $5 billion during FY 2010. The DOE requested $220 million for this activity during FY 2011.

The DOE is funding efforts to improve the efficiency of internal combustion engines for passenger and commercial vehicles while using advance fuel formulations that incorporate non-petroleum-based blending agents (biofuels) to reduce petroleum dependence. The effort should be redirected to eliminate the use of biofuel/gasoline mixtures and continue similar efforts using hydrogen gas as a fuel.

All federal support for biofuels should be eliminated by FY 2012. At least $3 billion can be saved during FY 2011 and $2 billion can be saved during FY 2012 by phasing out federal support.

2.4 Hydropower, Ocean Tides, and Waves

Department of Energy research and development activities that support hydropower, ocean tides, and waves should be eliminated for FY 2011 and beyond. The DOE has received appropriations of about $120 million from the beginning of FY 2009 to the end of FY 2010. More than $40 million can be saved by eliminating the FY 2011 Department of Energy congressional request for water power.

2.5 Clean Coal

Our coal-fired electric power plants provide about 48 percent of the total production of electric energy. Coal plants, automobiles, and trucks are the primary contributors to air pollution. To reduce the pollution caused by coal-fired electric plants, the DOE is funding efforts to extract carbon dioxide from the exhaust emissions—a result of coal combustion.

Investigations are underway to determine the feasibility of storing large volumes of carbon dioxide underground. The DOE budget for 2009 was allocated $2.4 billion for efforts that could lead to the safe storage of large amounts of carbon dioxide. Activity is underway to experimentally store carbon dioxide at many underground sites spread across our country. There is concern that government tests will be considered complete at the end of a relatively short time span. Can the DOE adequately determine the safety of storing carbon dioxide underground, for example, in less than ten years? If several sites used to store carbon dioxide begin to leak, our government assumes the liability for cleanup, using our tax money to correct the situation.

To safely store carbon dioxide underground, the gas must remain underground for hundreds of years without leaking into the atmosphere. Large volumes of naturally-formed carbon dioxide gas have been found beneath freshwater lakes. In one instance the carbon dioxide leaked to the surface of the lake and killed hundreds of people and animals living on land near the lake.

The common-sense approach to carbon sequestration is to pressurize and inject liquid carbon dioxide underground at one remote uninhabited location selected as safe and suitable for carbon sequestration. The ground area should be fenced off and posted as restricted against trespassing by humans. The area should be suitably instrumented and monitored to detect any emissions or leakage of carbon dioxide for a period of 100 years. If the test area proves to be safe and the 100-year test period is complete, possibly additional storage sites can be considered. If carbon dioxide underground sequestration is allowed to continue as planned before the completion of a 100-year field test, a near future financial government bailout for carbon dioxide cleanup is in the making. Remember, we are still paying for defense environmental cleanup. From FY 2008 to the end of FY 2010 the DOE has received appropriations totaling about $21.6 billion for defense environmental cleanup.

Since FY 2008, we have allowed our government to spend huge amounts of our tax money for fossil energy research and development when compared to similar expenditures for renewable energy. The DOE appropriation for FY 2008 and FY 2009 is $727 million and $863 million respectively. The American Recovery and Reinvestment Act 2009 allocated $3.4 billion for fossil energy research and development. The funding allocated for fossil energy R&D includes the funds allocated for coal R&D. The DOE allocation for FY 2010 is $672 million and the request for FY 2011 is $586 million.

Planning for, and the construction of, new coal-fired electric plants should be terminated in favor of planning for and installing natural-gas-fired, nuclear electric/hydrogen, and geothermal electric power plants over the next twenty years. The DOE funding for this activity in FY 2011 should be terminated for a savings of at least $586 million.

2.6 Energy Efficiency

Building Technologies

About $107 million in FY 2008 and $138 million in FY 2009 was appropriated for the Building Technologies Program. About $222 million was appropriated in FY 2010. The 2011 budget requested a little over $230 million for the Buildings Technologies Program. In addition to the $138 million for the FY 2009 appropriation, about $319 million was allocated from the American Recovery and Reinvestment Act 2009 for the Buildings Technologies Program. By the end of FY 2011, we will have invested over $1.0 billion since 2008 for the Building Technologies Program.

Residential and Commercial Building Integration

The DOE requested $78 million for this subprogram for FY 2011. Residential and commercial buildings primarily use electricity for air conditioning and lighting. Most residential and commercial buildings use natural gas or electricity for heating. A reduction in their use of electricity or natural gas does not reduce our need to import oil.

The subprogram should be terminated for FY 2011 because the effort does not provide a strong path to energy independence. Termination of this effort will result in a $78 million saving.

Emerging Technologies

The emerging technologies subprogram provides technology transfer information for equipment manufacturers that include energy efficient building materials such as windows, lighting and solar heating and cooling equipment. The DOE requested $92 million for FY 2011 to continue this subprogram. Funding for this activity should be reduced for FY 2011 by $50 million thereby bringing the activity back to the funding level of $42 million during FY 2009.

Technology Validation and Market Introduction

The purpose of the block grant program is to provide funding to units of local and state governments, and Indian tribes and territories to develop and implement projects to improve energy efficiency. By mid 2009, the DOE awarded more than 1,000 Energy Efficiency and Conservation Block Grants totaling over $1 billion of the total $2.7 billion available by the American Recovery and Reinvestment Act 2009. A key goal of the program is to demonstrate that when sufficient federally funded building modifications are reached within a community, non-participants will be convinced to invest in energy efficient modifications to their own buildings and homes.

The decision to invest in energy-efficient building modifications is usually predicated on the ability of the building owner to realize a reduction in electric or natural gas utility costs. Building modification costs have to be weighed against the savings realized by a reduction in utility costs. If the DOE is allowed to continue with the energy plan supported by the Obama administration, utility costs will increase and the justification for investing in higher-energy-efficient buildings will increase accordingly. However, if the cost of energy remains at the 2010 level or below, the incentive for investment in energy efficient building modifications becomes a mute subject. The Common Sense Energy Plan provides a pathway that will allow energy costs to remain at 2010 levels or less.

Why not motivate home and business owners to use the federal and state building codes to determine the economic common sense of making energy-efficient modifications to their buildings and homes? Taxpayers should not bear the expense of modifying a few homes and buildings with the hope that this activity may cause others to make similar private investments for the purpose of conserving energy. The DOE expenditures for private building and home modifications must be eliminated for FY 2011. Termination of this subprogram in FY 2011 will yield a $20 million saving.

Equipment Standards and Analysis Program

The Energy Policy Act 2005 and the Energy Independence and Security Act 2007 (that word, "independence," keeps repeating) authorizes the DOE to take steps to implement legally required energy-efficiency standards to meet the judicial and statutory deadlines. The Equipment Standards and Analysis program continues making rules affecting the design and manufacture of a variety of commercial products. It is disturbing to learn that the DOE responds to waiver requests from manufacturers and requests from manufacturers for information and recommendations. The DOE Office of Hearings and Appeals is responsible for providing information on their rules to manufacturers. It appears that this government agency may be overstepping their bounds into the free enterprise system. Many consumers select commercial products, after reviewing various consumer reports. Typically these reports conduct independent tests and comparative evaluations on a large variety of consumer products. By setting government standards on manufactured products, this agency may, inadvertently, be causing an increase in the price, to consumers, of some manufactured products.

The DOE Office of Hearings and Appeals should be eliminated. The Energy Policy Act 2005 and the Energy Independence and Security Act 2007 should be amended to scale down and phase out this DOE program.

The DOE requested $40 million for this subprogram during FY 2011. From FY 2009 to the end of FY 2010, DOE has received a total appropriation of $70 million. The subprogram should be scaled back to a technical assistance activity rather than an enforcer of energy efficiency standards. The FY 2011 funding request should be reduced from $40 million to the 2009 level of $20 million, which will result in a $20 million saving.

Industrial Technologies

Congressional funding in FY 2008 was $52.2 million for the DOE Industrial Technology program. Appropriated funding for FY 2009 was $88.2 million with an additional $261 million allocated from the American Recovery and Reinvestment Act 2009. Appropriations for FY 2010 were set at $96 million and the request for FY 2011 is $100 million. From 2008 to the end of FY 2011, our tax money supported over a half billion dollars.

Industries of the Future (Specific)

The Industries of the Future subprogram supports cost-shared development and demonstration of advance technologies to improve the energy and environmental performance of America's industries. In FY 2011, new work will be initiated with chemicals and cement industries, while existing efforts in the forest and paper products industry, the steel industry, and the aluminum industry will continue to completion. About $2.6 million is requested by the DOE for FY 2011. As this activity moves to completion, the DOE plans to shift to greater support of the cross-cutting technologies. This effort should be funded for FY 2011.

Industries of the Future (Cross-cutting)

The effort to develop and adopt alternate fuel and feedstock (biofuels) technology should be terminated in keeping with the Common

Sense Energy Plan to phase out all biofuel government support by 2012. The development of dual alternative fuels for turbines and engines that meet the stringent gas emissions regulations set in Southern California can be eliminated. With the introduction of electric and hydrogen-fueled cars and trucks, these activities are not productive and are a waste of taxpayer money.

The DOE requested $87 million for FY 2011. With the elimination of biofuel-related R&D, the funding request for FY 2011 should be reduced from $55 to $30 million, thereby saving $25 million. Congress has appropriated for this subprogram a total of $ 309 million from the beginning of FY 2009 to the end of FY 2010.

2.7 Eliminate Fuel-Cell Effort Related to Automobile Transportation

Fuel cells are not sufficiently developed for high production. They are unreliable and uneconomical. Our auto industry is equipped to manufacture internal combustion engines at high production rates. New engines can easily be designed and manufactured to burn hydrogen instead of gasoline or biofuels.

2.8 Stop New Interstate Highway Construction

In 1956, Congress created the Highway Trust Fund to finance the construction of the Interstate Highway System. The fund receives tax revenues from the federal gasoline tax of 18.4 cents per gallon. The fund is heading for bankruptcy as a result of expenditures for projects other than interstate highway construction. Expenditures from the fund should be used only to repair and maintain our existing interstate highways. With the introduction of high-speed passenger train transportation and the modernization of freight train transportation, there is no need to build new interstate highways.

2.9 Reduce Federal and Federally Funded Contractor Employment

The DOE Program Direction activity was allocated $127 million for FY 2009. For FY 2010, the allocation was $220 million (includes $40 million allocation from the Recovery Act, 2009). The DOE requested $200 million for FY 2011. The funds are being used to hire additional federal employees and provide for cost-of-living increases, in-grade increases, and the government's share of personnel benefits. The funding also provides for the hiring of additional support contractor staff, additional workspace, and the corresponding support systems required for new staff. In FY 2010, the DOE hired 253 additional federal employees. For FY 2011, the DOE has requested authorization to hire 392 additional federal employees. The DOE request for $200 million in FY 2011 should be eliminated and the hiring of new federal and contractor employees curtailed.

2.10 Summary of Funding, Tax Incentives, and Subsidy Reductions for FY2011 and FY 2012

Part of the Common Sense Energy Plan calls for reducing federal spending for our energy program. The table below provides a summary of the potential savings that can be realized by eliminating some programs and reducing the funding for others. The dollar amounts shown are in the thousands of dollars. For example, if federal subsidies, tax incentives, and DOE funding is eliminated for wind energy in FY 2011, the government could realize a savings of $1.5 billion. The DOE funding estimates for FY 2011 was taken from the Department of Energy FY 2011 Congressional Budget Request.

Energy Programs	FY 2011	FY 2012
Eliminate Wind Energy	$1,500,000	
Phase out Biofuels	$3,000,000	$2,000,000
Eliminate Hydropower, Ocean Tides and Waves	$40,000	
Eliminate Fossil Energy Research and Development	$586,000	
Building Technologies (funding reduction)	$186,000	
Industrial Technologies (funding reduction)	$25,000	
Eliminate New Federal and Contractor Employment	$200,000	
Total (dollars in thousands)	$5,519,000	$2,000,000

3.0 Reduce Air, Land, and Water Pollution

Introduction

The debate continues as to whether or not man-made air pollutants will cause a worldwide increase in climatic temperatures. We are witnessing a decrease in the size of the ice cap at the North Pole. As the ice cap melts, there is concern that ocean levels may rise and flood low-level private, commercial, and state-owned coastal properties. The good thing about the debate is that it has stimulated our interest to find clean sources of energy production. The debate has also reminded us of our objective in 1975 to eliminate our dependence on foreign oil and natural gas imports.

3.1 Clean Energy Production

The Department of Energy's Assistant Secretary Spurgeon reported in 2008 that no serious person can look at the challenge of maintaining our national security, reducing greenhouse gases, and addressing climate change and not come to the conclusion that nuclear power has to play a significant and growing role.

About thirty new nuclear electric power plants are being planned and twenty-one site locations have been established for construction. We have the technology readiness today to start building a much larger number of nuclear electric power plants that are needed. Both nuclear and geothermal electric power plants provide "green power" and produce no adverse air pollution. New nuclear/hydrogen production plants are being developed by the DOE and offer clean energy conversion of water to hydrogen and oxygen gas.

Natural-gas-fired combined-cycle electric power plants emit carbon dioxide as part of the combustion exhaust products. The gas emissions are significantly lower than the gas emissions resulting from the combustion of coal. The electric/hydrogen economy provides for the virtual elimination of gasoline and diesel-fueled automobiles and trucks. Such a reduction in automotive pollution will more than negate the added pollution from new gas-fired electric power plants.

An electric energy production mix of nuclear, natural gas, and geothermal provides our best choice for clean air and reduces the risk of land and water pollution. The mix of nuclear, natural gas and geothermal will provide the same degree of redundancy that nuclear, natural gas, and coal provides now.

3.2 Abundant and Economical Energy

The selection of an electric energy production mix of nuclear, natural gas, and geothermal assures an abundant and economic supply of energy now. We don't have to continue to pay for further research and development of biofuels and solar energy or spend our tax money for more wind turbines.

The new nuclear plants will use radioactive waste elements for fuel. The radioactive waste elements are generated by our nuclear plants now in operation. The radioactive waste has been accumulating since the 1960s and is being stored for future use. Our geologists predict that at the 1998 level of energy use, our natural gas supply will last about 200 years.

According to some electric industry estimates, the cost for generating electricity from a nuclear plant is about 0.4 cents per kilowatt-hour and about 7 cents per kilowatt-hour from a natural gas plant.

3.3 Electric, Hybrid, and Hydrogen-Fueled Passenger and Commercial Vehicles Operate Pollution Free

The electric/hydrogen economy, if advocated by our administration, can eliminate oil imports within a shorter time span than any other energy choice or group of renewable energy selections. The electric/hydrogen economy offers the best opportunity to reduce pollution and fossil fuel consumption by changing the fuel used for land-based transportation from gasoline to electricity and hydrogen gas.

Recall earlier that we have to import electric power from Mexico and Canada to meet peak demands. What will happen by 2015 if 500,000 potential electric car owners living in Los Angeles decide to recharge their car batteries all at the same time? Will we have increased our electric generating capacity to adequately meet the demand in Los Angeles or other large cities that have similar demands? By 2015, the majority of electric car owners will be unable to recharge their vehicles. Do we want to buy more electric energy from Canada and Mexico? Or should we build more of our own electric power plants to meet future demands?

The DOE is conducting a vehicle technologies program that supports research and development to make passenger and commercial vehicles more efficient and capable of operating on non-petroleum fuels. The non-petroleum fuels are, in general, biofuels. Although the program only provides limited activity with hydrogen, a non-petroleum fuel, the program strategies are designed to lead to environmental benefits, reduce oil use, improve America's energy security, and benefit the economy. The program correctly focuses on technologies for transportation electrification, including advanced batteries, power electronics, and electric motors for hybrid and plug-in hybrid vehicles. The plug-in hybrids can be recharged from an electric outlet or operated on liquid biofuels.

The DOE Vehicle Technologies program requested $325 million for FY 2011. From the beginning of FY 2009 to the end of FY 2010, the Vehicle Technologies program was allocated a total of $687 million. Of the $687 million amount, the American Recovery Act 2009 appropriated $109 million.

The DOE FY 2011 budget request significantly increases the emphasis on technologies that facilitate cost-effective plug-in hybrid vehicles and on the deployment activities to develop infrastructure for transportation electrification and to accelerate the use of maturing technologies such as alternate fuels. However, only one alternate fuel—hydrogen—should be considered for the vehicle technology effort.

3.4 Research New Energy Sources

One of the most promising sources for renewable energy is the fusion process. The challenge is to be able to extract more energy from the fusion process than is required to operate a fusion reactor.

The goal of the International Thermonuclear Experimental Reactor is to produce ten times more energy than is required to operate it. The fusion reactor offers a clean source for energy without producing waste materials containing radioactive elements. Our country should fully support the scientific and engineering efforts to complete construction of the experimental reactor. Construction of the reactor is scheduled for completion in 2016. We should be fully involved with the experiments planned for the reactor when it is ready for operation.

If the experiments prove to be successful, the next step will be to build a fusion reactor designed to produce electricity. By 2040, the nuclear fusion electric power plant may well begin to replace the majority of our electric power plants operating today.

4.0 Reduce Consumption of Fossil Fuels

Introduction

Part of the changes needed in our energy policy, to reduce foreign oil imports, is to implement methods for reducing our current consumption of fossil fuels. Fuel conservation will help to propel our country toward the goal of energy independence. For example, interstate highway speed limits can be reduced. A reduction in Interstate speed limits can be implemented in a matter of a few months. By modernizing our passenger and freight rail transportation system, we can substantially reduce our consumption of fossil fuels. In addition to fuel savings a substantial decrease in highway deaths and injuries will be realized by transferring highway transportation to high speed rail and modern rail freight carriers.

Highway safety information is gathered by local, state, and federal government. Conclusions drawn from this information indicates at least two significant methods for reducing highway accidents. First, year-to-year comparisons show highway accidents diminish when fewer trucks and automobiles travel our highways. Second, it has been concluded that enforced and reduced highway speed limits, when compared from one year to the next, have resulted in a reduction of auto-related deaths and injuries. We should seriously consider ways to reduce the frightening number of our people that are killed or injured as a result of auto-related accidents.

Brad Linscott

The Common Sense Energy Plan calls for building a high-speed passenger train system that will in effect remove thousands of cars from our highways. The plan calls for modernizing our rail freight system to attract more commercial use including our highway freight companies. A modern rail freight system will reduce the number of tractor trailer rigs traveling our highways.

For those lonely travelers remaining on our highways, modern equipment is needed to control automobile and truck speeds. Not only will human fatalities and injuries be reduced, we will also be rewarded with a decrease in our auto insurance rates and a reduction in our nationwide medical expenses.

4.1 High-Speed Passenger Trains

To help reduce our consumption of fossil fuels we need to design, produce, and put into service passenger trains capable of speeds between 150 and 200 miles per hour. High-speed train routes should be selected to provide direct competition with "short hop" air transportation.

A plan for high-speed passenger train service between large cities less than 500 miles apart should be initiated this year. The plan should be implemented in FY 2012 to begin rail bed construction, procurement of high-speed trains, and construction of modern terminals for travelers. Trains can use hydrogen to fuel internal combustion engines instead of the diesel engine used to drive electric generators. The electric generators supply power to electric motors used to drive the train. Passenger trains will in effect reduce the number people traveling on our highways. A reduction in the number cars on our highways we will result in a reduction of highway accidents.

4.2 Interstate Speed Control

To help reduce our consumption of imported fossil fuels and increase driver safety, speed limits on federal and state interstate highways must be reduced to a range between 55 and 60 miles per hour.

To enforce newly imposed speed limits on interstate highways, automated speed enforcement similar to that used in France is needed. The automatic speed enforcement system provides a new age in public road safety policy. Federal and state governments should provide incentives and funding to support the modernization of interstate highway traffic control, instead of spending our money for bicycle trails and pedestrian sidewalks.

4.3 City and Urban Traffic Control

A centralized video-monitored computer traffic-flow and speed control system is needed in our cities and heavily populated suburbs. Heavily traveled highways need traffic light systems that are synchronized to allow a smooth flow of traffic. On demand entry from secondary streets and shopping malls should be eliminated during rush-hour traffic. Video monitors with the capability to issue traffic tickets are needed at traffic intersections. For the case of traffic tie-ups due to accidents, the centralized video monitoring system should have the capability to transmit this information to automobiles with navigation equipment traveling in the vicinity of the traffic jam.

City and urban traffic control systems offer the opportunity to reduce fuel consumption and human injury resulting from accidental car collisions. A reduction in auto-accident-related human injuries will enable our insurance companies to lower their insurance rates and decrease the number of people needing medical assistance.

5.0 Organize to Implement the Plan

In his letter to me in May of 2009, the United States senator from Ohio, George Voinovich, stated that, "It is long past time for the United States to undertake an Apollo-like program to wean ourselves away from oil dependence and on to clean, reliable, and domestically abundant energy alternatives."

The space program during the 1960s provides the management guidelines needed for organizing, planning, leading, and controlling a nationwide energy program. It is suggested that management methods similar to those for the space program be used to manage the implementation of the Common Sense Energy Plan. The Common Sense Energy Plan can be considered the beginning of America's Second Declaration of Independence, which, when implemented, will harmonize our energy, environmental, and economic policies.

The DOE should be assigned as the lead agency and organized to carry out the Common Sense Energy Plan. The Department of Transportation should be the lead agency and be organized to carry out efforts to introduce high-speed passenger service between major cities across our country. They need to assist our auto industries to make the transition from fossil/biofuels as an energy source to energy supplied by electricity and hydrogen gas. The Department of Transportation needs to lead the modernization of our rail freight system and help to attract our highway freight transport industries by offering them economical rail transportation and high-speed delivery as an alternative to using the interstate highways.

References

1. Letter from George V. Voinovich, United States Senator, Ohio, to Bradford S. Linscott. Subject: America's Energy Policies. May 1, 2009.

2. "Emissions of Greenhouse Gases Report." DOE/EIA-0573 (2007). December 3, 2008.

3. Dooley, J.J. "U.S. Federal Investments in Energy R&D: 1961–2008." DOE/PNNL-17952, October 2008.

4. "Analysis of the Proposed Reorganization of the Department of Energy." Congressional Budget Office. February 3, 1982.

5. Wiser, Bryan, et.al. "Using the Federal Production Tax Credit to Build a Durable Market for Wind Power in the United States." Ernest Orlando Lawrence Berkley National Laboratory, Environmental Energy Technologies Division. LBNL-6358. November, 2007.

6. Linscott, Bradford S. "Large Horizontal-Axis Wind Turbines." DOE/NASA/20320–58 March, 1984.

7. Fingersh, L. "Wind Turbine Design Cost and Scaling Model." Technical Report. NREL/TP-500–40566. National Renewable Energy Laboratory. December, 2006.

8. Porretto, John. "Pickens Wants to Unload Hundreds of Wind Turbines." Associated Press. Plain Dealer. July 8, 2009.

9. "Electric Power Annual, Electric Power Industry 2007: Year in Review." Energy Information Administration. January 21, 2009.

10. Service, Robert F. "Is It Time to Shoot for the Sun?" *Science.* Vol. 309. July 22, 2005.

11. Fisher, Daniel. "A Brief History of Energy Boondoggles." *Forbes Magazine*, Special Issue. November 24, 2008. p.82.

12. "Energy Savers, Ocean Tidal Power." U.S. Department of Energy. <www.energysavers.gov/renewableenergy/ocean/index.cfm/myoptic>

13. "Nova Scotia Renewable Public Education in Tidal Energy." Nova Scotia Department of Energy. June 6, 2009. <www.gov.ns.ca/energy/renewables/public- education/tidal.asp>

14. "Ocean Wave Energy." OCS Alternative Energy and Alternate Use. <http://ocsenergy.anl.gov/guide/wave/index.cfm>

15. "National Biofuels Action Plan." Biomass Research and Development Board. October, 2008.

16. "Corn-for-ethanol's Carbon Footprint Critiqued." *Science Daily.* March 11, 2009.

17. Duchane, D. and Brown D. "Hot Dry Rock (HDR) Geothermal Energy Research and Development at Fenton

Hill, New Mexico." Los Alamos National Laboratory, Los Alamos, NM, GTHC Bulletin. Dec. 2002.

18. Kutscher, Charles F. "The Status and Future of Geothermal Electric Power." NREL/CP-550–28204. August, 2000.

19. "FY 2010 Congressional Budget Request." U.S. Department of Energy. DOE/CF- 041. Volume 7. May, 2009.

20. "Building Energy Codes Program." About the Program, U.S. Department of Energy. <http://www.energycodes.gov/what-wedo/index.stm>.

21. "Energy Efficiency and Conservation Block Grant Program." U.S. Department of Energy, Energy Efficiency, and Renewable Energy. 2009.<http://eecb.energy.gov>

22. "DOE Request for Information." Recovery Act: Energy Efficiency and Conservation Block Grant Program: Competitive Grants, In support of the American Recovery and Reinvestment Act of 2009. September 14, 2009.

23. Mufson, Steven. "Pipe Dream or Clean Coal Plant in Effort to Slow Down Climate Change." *Plain Dealer Newspaper.* August 16, 2009.

24. "FutureGen Clean Coal Project." U.S. Department of Energy. July, 2009. <http://fossil.energy.gov/programs/powersystems/futuregen/index.html>

25. "Poisonous Gas from Lake Threatens 2 Million People." *Plain Dealer Newspaper.* August 15, 2009.

26. "FY 2010 Congressional Budget Request." U.S. Department of Energy. DOE/CF- 037. Volume 3. May, 2009.

27. "Electric Generation Using Natural Gas." <http://www.naturagas.org/overview>

28. "Natural Gas Imports and Exports." First Quarter Report 2009, Office of Natural Gas Regulatory Activities, Office of Fossil Energy, U.S. Department of Energy.

29. "Natural Gas Consumption by End Use." Energy Information Administration. Official Energy Statistics from the U.S. Government.

30. "The Pressure Builds." *The Wall Street Journal.* June 30, 2008. p. R11.

31. Hurdle, Jon, "Pennsylvania says natgas drilling risks inevitable." March 20, 2009 <Reuters.com>

32. Gashler, Krisy. "Experts question treated gas-drilling fluid." July 9, 2009 < Ithacajournal.com>

33. "Nuclear Power 2010." <www.ne.doe.gov.np2010/neNP2010a.html>

34. "Nuclear Power 2010." Nuclear Power Deployment Scorecard. May 11, 2009.

35. Stone, Andy. "An Interstate Highway System for Energy." *Forbes Magazine Special Issue, Energy & Genius.* November 24, 2008. p. 108

36. "Energy & Genius." *Forbes Magazine, Special Issue.* November 24, 2008. p. 130.

37. "Hydrogen Fact Sheet, Hydrogen Production-Nuclear." New York State Energy Research and Development Authority.<www.nyserda.org>

38. "Hydrogen Posture Plan, An Integrated Research, Development and Demonstration Plan." U.S. Department of Energy and U.S. Department of Transportation. December, 2006.

39. "Nuclear Hydrogen Initiative." DOE-Office of Nuclear Energy. <http://nuclear.energy.gov/NHI/neNHI.html>

40. "Highway Trust Fund Expenditures on Purposes Other than Construction and Maintenance of Highways and Bridges during Fiscal Year 2004–2006." United States Government Accountability Office. June 30, 2009.

41. 41 "The Broken Highway Trust Fund." Republican Caucus, The Committee on the Budget. July 23, 2008.

42. "Fatality Analysis Reporting System." National Statistics.< http://www-fars.nhtsa.dot.gov/Main>

43. "National Traffic Fatalities On the Decline." The National Safety Commission Alerts. August 6, 2008 <www.nationalsafetycommision.com>

44. Coston, James "The Myth of Passenger Train Profitability." Speech by member of Amtrak Reform Council, Philadelphia. December 1, 2001.

45. Krigman, Eliza. "High-Speed Trains are fast and clean and you can put them under the English Channel." *American Chronicle.* November 27, 2008. www.americanchronicle.com/articles.

46. Marshall, A. "Backers of Train Service Cheer Project's Launch." *The Plain Dealer Newspaper.* January, 29, 2010.

47. "The Most Dangerous Intersection In Ohio." *The Plain Dealer Newspaper.* October 9, 2009.

48. McCarty, James F. "Lawyers loophole beats traffic cameras." *The Plain Dealer Newspaper.* p.1 February 20 2009

49. "Energy & Genius." *Forbes Magazine, Special Issue.* November 24, 2008. p. 126.

50. "Use of Hydrogen Grows to Fuel Vehicles, Produce Electricity." Telling America's Story, U.S. Department of State. February 25, 2008. <America.gov>

51. "Effects of a Transition to a Hydrogen Economy on the Employment in the United States Report to Congress." U.S. Department of energy. July, 2008.

52. "FY 2008 Progress Report for Advanced Combustion Engine Technologies energy Efficiency and Renewable Energy." Office of Vehicle Technologies, U.S. Department of Energy. December, 2008.

53. Mangels, John. "U.S. support of reactor has been inconsistent." *The Plain Dealer Newspaper.* January 21, 2009.

54. Fahey, Jonathan. "Star Struck." *Forbes Magazine, Special Issue: energy & Genius.* November 24, 2008. p.114.

55. "The U.S. Generation IV Fast Reactor Strategy. " U.S. Department of Energy Office of Nuclear Energy. DOE/NE-0130. December, 2006.

56. Eaton, Sabrina. "Senator Sherrod Brown wants American Parts in U.S. Funded Wind Turbines." *The Plain Dealer Newspaper.* March 5, 2010.

57. "Department of Energy FY 2011 Congressional Budget Request." DOE/CF-0049. Volume 3. February, 2010.

listen|imagine|view|experience

AUDIO BOOK DOWNLOAD INCLUDED WITH THIS BOOK!

In your hands you hold a complete digital entertainment package. Besides purchasing the paper version of this book, this book includes a free download of the audio version of this book. Simply use the code listed below when visiting our website. Once downloaded to your computer, you can listen to the book through your computer's speakers, burn it to an audio CD or save the file to your portable music device (such as Apple's popular iPod) and listen on the go!

How to get your free audio book digital download:

1. Visit www.tatepublishing.com and click on the e|LIVE logo on the home page.
2. Enter the following coupon code:
   1ef2-8a6d-5a3e-de20-5103-1b85-faae-cac7
3. Download the audio book from your e|LIVE digital locker and begin enjoying your new digital entertainment package today!